Hot-Dip Galvanizing of Steel Structures

Hot-Dip Galvanizing of Steel Structures

Vlastimil Kuklík

Jan Kudláček

ELSEVIER

AMSTERDAM • BOSTON • HEIDELBERG • LONDON
NEW YORK • OXFORD • PARIS • SAN DIEGO
SAN FRANCISCO • SINGAPORE • SYDNEY • TOKYO
Butterworth-Heinemann is an imprint of Elsevier

BH

Butterworth-Heinemann is an imprint of Elsevier
The Boulevard, Langford Lane, Kidlington, Oxford OX5 1GB, UK
50 Hampshire Street, 5th Floor, Cambridge, MA 02139, USA

ISBN: 978-0-08-100753-2

British Library Cataloguing-in-Publication Data
A catalogue record for this book is available from the British Library.

Library of Congress Cataloging-in-Publication Data
A catalog record for this book is available from the Library of Congress.

For Information on all Butterworth-Heinemann publications
visit our website at http://store.elsevier.com/

Contents

Preface

The authors were compelled to write this monograph because, although many publications exist on the production of hot-dip galvanized coatings, there are few that provide a comprehensive overview of information about batch hot-dip galvanizing technology that is useful not only for galvanizing plant personnel but also for users of this anti-corrosion protection method for steel.

The contents of this book expand the state of knowledge in the hot-dip galvanizing field using the latest research results that are related to hot-dip galvanizing technology, mechanisms of formation of zinc-based alloy coatings and their accompanying phenomena, and last, but not least, the integrity of hot-dip galvanized steel structures.

The book offers a summary of practical findings about the batch hot-dip galvanizing process. The text also partly deals with the issue of steel corrosion and the principle of zinc-based anti-corrosion protection and provides a brief overview of the most frequently used zinc coating methods. The publication mainly focuses on the technology of applying galvanized coatings in commercial galvanizing plants. In its clearly arranged chapters it describes the principles of designing and producing parts to be hot-dip galvanized, the usual surface pre-treatment processes, the metallurgy of formation of alloy coatings, including their morphological variations, and coating defects as well as repair methods in a detailed way. The final chapters concentrate of the service life of zinc coatings, the European standards, and related legislation with an emphasis on environmental aspects of hot-dip galvanizing.

In this richly illustrated book, the authors have drawn mainly upon the latest global scientific research in the field as well as their own broad experience and numerous experiments conducted to supplement, extend, and verify the conclusions. They hereby want to offer personnel of commercial galvanizing plants a suitable guide that will help them find their way in the field of hot-dip galvanizing as well as related areas. The guide may also provide useful information to other interested people — from students in specialist schools and universities to investors, designers, and manufacturers of zinc-coated parts, corrosion engineers, or surface finishing inspectors.

Some texts cited in the original (2014) Czech language edition have been supplemented and modified for this English language edition.

The original book was reviewed by Viktor Kreibich, Head of the Institute of Engineering Technology on the Faculty of Mechanical Engineering of the Czech Technical University in Prague, and this English translation has been assisted by Murray Cook, Executive Director of European General Galvanizers Association.

The authors thank the reviewers for their constructive comments that contributed to the achievement of the required professional standard of the published work and also thank the authors of the adopted expert articles that supplement the original texts in the required scope.

New York 2015

Introduction

The method of zinc coating of ferrous metal items by dipping in molten zinc to protect them from corrosion was first officially demonstrated by the French physician Paul Jacques Malouin in 1742 (Figure I.1) [2]. In 1799, Alessandro Volta (Figure I.3) [3] contributed to knowledge of the principle of cathodic protection of iron from corrosion by discovering the galvanic cell. He derived its name from the Italian physician Luigi Galvani (Figure I.2) [4], who observed the effects of electric current and described them in 1791 in his Treatise on Electric Forces. The principle of cathodic protection of 1829 was explained by Michael Faraday (Figure I.4) [5] through an experiment with zinc and iron nails immersed in salt water.

It took nearly a hundred years from the experiments conducted by Malouin for the French chemist Stanislas Sorel [6] to file a patent for galvanic protection of iron from corrosion by application of zinc coating by immersion of a chemically pretreated zinc object in molten zinc. The patent was filed in 1837 and he called his method galvanizing. Shortly after that, around 1840, the first factories were established in France, Great Britain, and Germany where iron objects were industrially coated with zinc (Figure I.5).

Figure I.1 Paul J. Malouin [2].

Figure I.2 L. Galvani [3].

Figure I.3 A. Volta [4].

Figure I.4 M. Faraday [5].

Figure I.5 From the history of hot-dip galvanizing [7].

I.1 Zinc

Zinc is a low melting point, transition metal of the twelfth group of the periodic table of elements. It has a silverish blue and white color with distinctive gloss. It has a hexagonal crystal structure with the tightest arrangement (Figure I.7) and it exhibits considerable anisotropy. At normal temperatures it is tough but at slightly elevated temperatures above 100°C it is very easy to form. At low temperatures it becomes brittle. It exhibits very low strength and hardness. In Earth's crust and sea water, where it is only found in the form of compounds, it is present in a relatively high concentration. The principal raw materials for its production are sulfide (Figure I.6) and oxide ores. From there it is isolated by roasting of enriched ore in

Figure I.6 Zinc ore sphalerite.

Figure I.7 Crystalline structure of zinc.

multiple-hearth or fluid smelters. Roasted ore containing zinc in oxide form is subsequently agglomerated and distilled or, through metallurgical processes, it is first transformed to the sulphate form by leaching, from which it is then separated by an electrolyte.

Pure (industrially produced) zinc quickly oxidizes when exposed to the air and is prone to corrosion. Under favorable atmospheric conditions it becomes covered by a consistent layer of insoluble zinc carbonate that is highly stable.

As a trace element, zinc is present in living organisms and in all parts of the human body (in organs, tissues, and bones). It is essential for organisms, especially for their growth and for reinforcement of the immune system. Zinc is non-toxic. Organisms ingest it together with food and excrete any excess quantities. It is also an essential raw material for the pharmaceutical industry.

Its very good casting properties are important in industrial uses; it is used to produce castings of complex shapes. Zinc also exhibits good alloying characteristics; it combines well with other metals. It is an important alloying metal for the production of brass. Zinc oxide is used for the production of zinc white and a considerable share of its production is consumed as a filler used in the rubber-making industry.

An important field of industrial use of zinc is the surface finishing of materials. Zinc coatings provide highly efficient and durable protection from corrosion for steel structures and components. The highest volume of zinc production is used for the corrosion protection of iron and steel. Table I.1 below presents an overview of zinc's characteristics.

Table I.1 Overview of zinc characteristics

Zinc		
chemical symbol	Zn	
atomic number	30	
atomic weight	65.37	
characteristic group	transition metals	
valence	2	
color shade	silverish bluish white, lustrous	
crystalline structure	hexagonal (with closest packing)	
density ρ_{20}	7140	$[kg.m^{-3}]$
melting temperature t_0	419.5	[°C]
boiling temperature t_1	907	[°C]
thermal expansion coefficient α_{20}	29.10^{-6}	$[K^{-1}]$
specific latent heat of melting l_t	102	$[kJ \cdot kg^{-1}]$
thermal conductivity λ_{20}	109	$[W \cdot m^{-1} \cdot K^{-1}]$
specific thermal capacity C	385	$[J \cdot kg^{-1} \cdot K^{-1}]$
tensile elasticity modulus E	100.10^3	[MPa]
torsion elasticity modulus E	40.10^3	[MPa]
tensile strength σ_t	120	[MPa]
ductility ψ	60	[%]

(Continued)

Table I.1 (Continued)

hardness	72	[HB]
electrical conductivity	$16{,}6.10^6$	$[S \cdot m^{-1}]$
specific electrical resistivity ρ	59	$[n\Omega \cdot m]$
electrode potential	−0.76	[V]
magnetic properties	diamagnetic	

I.2 Steel corrosion and corrosion protection

In general, corrosion is a very complex process. An understanding of processes and phenomena related to corrosion protection by zinc coatings and of the basic principles must be taken from the current state of knowledge of these aspects as described in the relevant specialized literature (e.g., [8,9]). To provide some guidance on these topics the following explanations give only basic information that is dealt with in more detail by other sources.

I.3 Steel corrosion

Corrosion of metals generally means their undesired, spontaneous, gradual, and irreversible destruction caused by physical-chemical or chemical action of the environment or by biological influences [10,11]. Steel corrosion (Figure I.8) is disruption of the metal due to its electrochemical or chemical reaction with a corrosive environment.

For chemical corrosion of metals (Figure I.9) the reaction occurs in electrically non-conductive, generally gaseous environments. On contact with metal the gas is first dissociated into ions that adsorb on the metal surface (Figure I.9A) and subsequently, due to their chemical reaction with the metal, corrosion products are generated that cover the substrate (Figure I.9B). The corrosion products of chemical corrosion are produced by a direct chemical reaction between the metal and the environment without participation of an electrolyte. A typical example of chemical corrosion is formation of scale during heat processing of scale. During chemical corrosion of steel, oxidation of the metal and reduction of the oxidation component occur at the same time.

From the point of view of the principles and technology of hot-dip zinc coating, it is important to know that steel corrosion is also caused by liquid metals. This is a special case of chemical corrosion where the structural metal does not oxidize but either dissolves in liquid metal or alloy phases are formed that lead to a loss of coherence of the structural metal [9].

Electrochemical reaction in a corrosion cell goes on in the presence of an electrolyte when ionization of the metal (release of cation) and reduction of the

Figure I.8 Steel attacked by electrochemical corrosion.

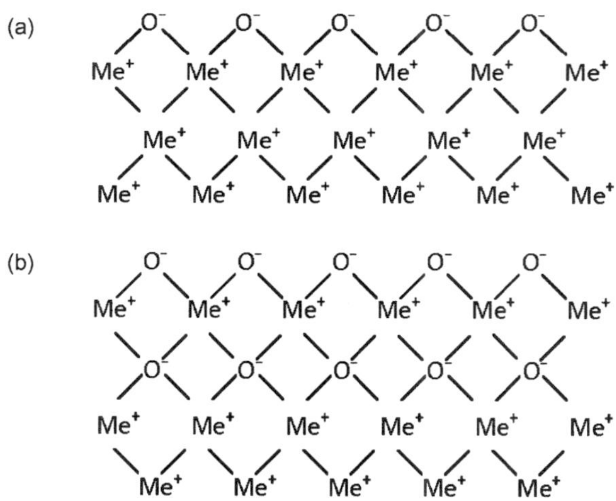

Figure I.9 Principle of chemical corrosion of metals.

oxidation component of the corrosion environment are not in mutual agreement and their rate depends on the electrode (electrochemical) potential of the metal. Corrosion rate is the quantity of metal converted per unit time to corrosion products related to a unit area [12].

The main cause of corrosion of steel exposed to the atmospheric environment is the air humidity, which is a condition for electrolyte formation. Its corrosivity depends on the type of polluting and other substances in the atmosphere and on their concentration. Air pollution by sulphides or chlorides plays an important role. In the industrial environment a determinant of ambient corrosivity is an increased concentration of sulfur dioxide, which is mainly the product of fossil fuel combustion. Chlorides may result from increased salinity of the atmosphere in seaside environments, but an important source is also salt spread on roads.

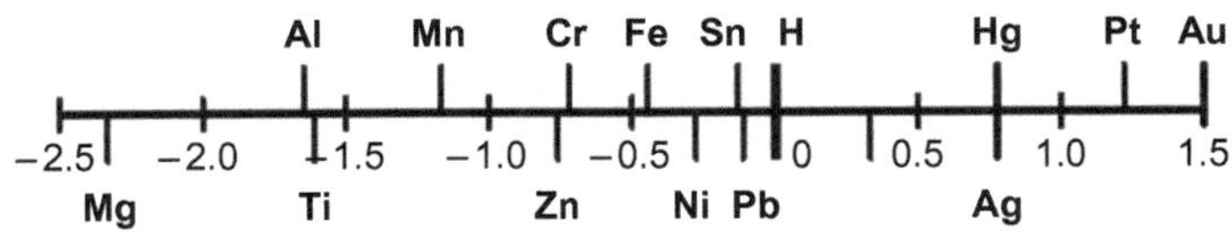

Figure I.10 Electrode potentials of selected metals in salt brine at 20°C [9].

A cause of electrochemical corrosion may consist of an electrically conductive connection between two metals with different electrode potentials in an electrolyte and/or the creation of a concentration cell in a non-homogeneous corrosive environment, e.g., due to variable access of atmospheric oxygen to the steel surface (the aggressiveness of the solution increases in places with limited oxygen access) [8].

Corrosion resistance of structural steel in the atmospheric environment and in water and soil is generally low. Steel is very quickly subject to corrosion in acidic solutions. Conversely, it resists corrosion relatively well in diluted alkaline solutions.

It is important to note that, in a bimetallic cell, it is always the metal representing the anode that corrodes as oxidation is the characteristic for the anode. The cathode is subject to reduction and the metal of the cathode does not corrode. At the same time there is a rule that the metal in the pair that has a lower electrode potential in the particular environment becomes the anode. This metal corrodes first while the cathode metal (with a higher electrode potential value) passes into the immune state.

Figure I.10 shows a list of electrode potentials (redox potentials) of some selected metals from the galvanic series in a diluted aqueous solution of NaCl at 20°C.

I.4 Anti-corrosion protection of steel using zinc

The way in which zinc protects steel from corrosion [13] results from its favorable physical and chemical characteristics, which are very suitable for this purpose (see also Chapter 8):

A. In an electrically conductive connection with iron, zinc behaves as a sacrificial anode under suitable conditions in most normal environments, i.e., the anode is first attacked by corrosion and ensures that iron is the cathode, which passes into the immune state and thus does not corrode.

B. A significant feature of zinc is that when it is deposited on the surface of ferrous metals it can create a highly resistant and adherent metallic barrier with the required thickness to separate the coated product from a corrosive environment.

C. An important aspect is the ability of zinc to be efficiently passivated. If the zinc surface is suitably treated its corrosion rate is decelerated. If a zinc coating is exposed in a dry and sufficiently ventilated environment it becomes covered by a consistent (continuous) layer of corrosion products with high mechanical and mechanical resistance (patina), which decelerates its corrosion in normal atmospheres.

List of selected zinc coating technologies

1

With regard to its properties, zinc has proven its quality as a coating material to protect ferrous metals from corrosion and, over the course of time, a number of technologies for its application and use in surface finishing have been developed.

1.1 Electroplating of zinc

A very common zinc coating technology is electrolytic deposition of a zinc coating from an electrolyte [14]. In this system, a zinc electrode (anode) is connected to the positive pole of a source of direct electric current (Figure 1.1). Zinc ions contained in the electrolyte are deposited on the negative cathode represented by the steel product that zinc ions adhere to, forming a zinc layer. A precondition for successful deposition of the coating is chemical and electrochemical pre-treatment ensuring a perfectly pure metallic surface of the parts is achieved. The thickness of the deposited zinc layer is generally 10 to 20 μm and it can be controlled very exactly. Besides pure zinc coatings, alloy coatings are used as well (most commonly zinc-nickel).

1.2 Thermal spraying of zinc (metallizing)

Thermal spraying (Figure 1.2) is used to apply zinc that is melted by flame or electric arc and carried by a gas stream to the sand-blasted surface of the part to be coated. The coating adheres by mechanical adhesion [15]. Pure zinc may be used for thermal spraying for anti-corrosion, but for higher corrosion resistance, alloys of zinc with aluminum are also frequently used. The coating is applied as a layer with the thickness of 80 to 250 μm.

For reliable adhesion of a zinc coating applied by metallizing the surface requires pre-treatment by sand-blasting with the use of sharp-edged particles.

After the application of sprayed zinc it is necessary to apply an organic painting material on the deposited coating that will adhere very well in its profile. Combination of zinc coating with organic paint (referred to as a duplex system, see Section 8.3) offers a very good corrosion resistance with regard to efficient synergy of both the materials.

Hot-Dip Galvanizing of Steel Structures. DOI: http://dx.doi.org/10.1016/B978-0-08-100753-2.00001-X

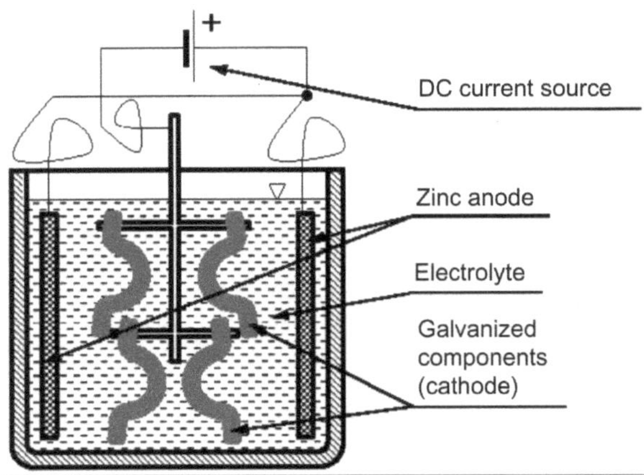

Figure 1.1 Electrolytic zinc plating principle.

Figure 1.2 A metallizing gun for thermal spraying in the powder version is suitable for repairs of galvanized coating applied by immersion in molten zinc.

1.3 Hot-dip galvanizing (by immersion in molten zinc)

Parts designed to be zinc coated are first chemically pre-treated to achieve a clean steel surface. Then the surface is activated with flux and subsequently the parts are immersed in a bath of molten zinc. During immersion in the bath, a zinc coating consisting of a few alloy layers is formed on the steel surface. The coating is formed depending on the instantaneous conditions of diffusion of iron and zinc atoms and their mutual bond. In the bath, a spontaneous reaction between iron and zinc occurs, producing several iron-zinc alloy phases.

Table 1.1 Comparison of hot-dip galvanized coatings applied by immersion in commercial zinc coating plants

Application method		Characteristic	Application
Batch	dry process	limited handling during coating application, high productivity	parts that can be suspended individually
	wet process	good handling during coating application, limited weight	complex shapes requiring manual handling
Centrifuged	low-temperature	the coating has the same characteristics as in the case of batch application, limited possibility to control the coating thickness	minor parts requiring a bigger coating thickness
	high-temperature	the coating lacks the alloy phase ζ, good possibility to control the coating thickness	minor parts requiring exact control of the coating thickness

Batch hot-dip galvanizing (commercial) is carried out as a dry (see Section 2.2.1) or wet process (see Section 2.2.2) or as coating with centrifuging (see Section 2.2.3). Given the focus of this book, in the chapters below we will mainly concentrate on these technologies. A basic comparison of individual galvanized coating types by the method of their application is presented in Table 1.1. Besides these technologies there are methods for continuous zinc coating of strips, wires, or pipes.

Table 1.1 provides a general overview of the characteristics and use of coatings produced in commercial hot-dip galvanizing plants. Galvanized layers applied by various immersion technologies have specific characteristics, which results in specific features of their practical use.

1.4 Sherardizing (diffusion zinc coating)

Sherardizing is a method of diffusion zinc coating application [16, 17] where small parts cleaned by sand-blasting (or exceptionally pickling) are tumbled in a rotary furnace together with zinc dust mixed with sand. The temperature of the charge is just below the melting point of zinc. During the process, zinc atoms diffuse into the steel surface and at the same time zinc may get chemically bonded to iron in the surface layer of steel. An advantage is relatively high corrosion resistance of the coating, maintaining of geometry of the zinc-coated part, and elimination of the hydrogen embrittlement risk. Figure 1.3 shows a comparison of coating applied by centrifuge hot-dip galvanizing (top) and by sherardizing (bottom).

Figure 1.3 Different appearance of the coating of a centrifugally galvanized (top) and sherardized screw (bottom).

1.5 Mechanical plating

Mechanical plating is carried out in a rotary steel tumbler with plastic lining at the ambient temperature. Pre-treated parts are tumbled for a few tens of minutes in an appropriate chemical environment together with zinc dust, glass balls, and a suitable promoter [15]. Supported by impacts of glass balls, zinc particles are cold welded to the material of the parts to be galvanized and zinc sticks to their surface. When the required coating thickness is achieved, the batch is removed from the tumbler, separated from the glass balls, rinsed, and usually dried in a centrifuge. Parts that are surface treated by mechanical plating do not face the risk of hydrogen embrittlement or loss of steel strength. The thickness of coating applied by mechanical galvanizing is up to 15 μm and is very homogeneous. This method is suitable for precise parts made of hardened or heat-treated steel, e.g., springs.

1.6 Zinc coating comparisons

Table 1.2 provides an overview and comparison of properties of zinc coating applied by different techniques. The described characteristics of individual coating types are only approximate; some parameters mentioned here may vary.

Table 1.2 Comparison of zinc coating types from the point of view of different technologies

Zinc coating type	Electroplated	Thermally sprayed	Hot-dip galvanizing	Sherardizing	Mechanical plating
Usual coating thickness [μm]	up to 20	80–250	35–250	10–45	up to 15
Mechanical resistance	good	good	very good for Si < 0.03 %, reduced for 0.03 < Si < 0.12 % and Si > 0.25 %	very good	reduced
Benefits	uniform appearance, preciseness	possibility to control thickness, without limitation of part dimensions, no risk of hydrogen embrittlement	metallurgical bond with the substrate, high productivity and good economy, coating of inner surfaces	maintaining the geometry, no risk of hydrogen embrittlement	maintaining the geometry, no risk of hydrogen embrittlement
Disadvantages	hydrogen embrittlement, problems with coating of inner surfaces	the coating is porous and requires application of sealer, coating on the outer surfaces only	heterogeneous appearance, high variation of coating thickness and roughness, change of geometry	low productivity, high energy demands	low adhesion, insufficient thickness for anti-corrosion protection
Application	fasteners, mechanical engineering applications	steel structures	wide application range from small parts to steel structures	fasteners for heavy-duty applications	for light-duty applications in interior situations
Appearance	light glossy to matte with a number of color shades depending on the subsequent conversion coating	light matte	light glossy to dark matte	dark matte	light matte

Hot-dip galvanizing

2

Hot-dip galvanizing is a process of creating an alloy coating of zinc on steel parts by their immersion in a zinc melt (see Chapter 4). This operation involves a complex series of diffusion processes, elementary metallurgical reactions, and thermodynamic transformations. In terms of appearance, thickness, structure, and other characteristics, the final coating is the result of the influence of a number of factors [18].

A precondition for successful galvanizing is perfect wetting of the zinc-coated part by the zinc melt. Therefore parts intended to be galvanized are subjected to chemical pre-treatment before galvanizing to achieve a pure metallic surface and, then, its activation. The parts are first degreased and then pickled. Pickling removes scale, oxides, and corrosion products and possibly other soluble impurities from the surface. Before immersion in molten zinc, a flux is applied onto the component by immersion in a flux bath and subsequent drying if the "dry process" is used, or in the "wet process" flux is spread onto a part of the zinc bath surface (separated by a slide).

In the **wet process**, parts to be galvanized are generally immersed in zinc manually with the use of tongs or hooks as each individual part immersed in the zinc bath through the flux layer must be removed from it through the other clean part of the bath surface.

In the **dry process**, flux is applied in advance by immersion of the entire batch of components in a flux bath and then it is dried (hence the name dry process) before the batch is immersed in the zinc bath. In dry process, handling of parts to be galvanized is mechanized in the galvanizing plant. The limiting factor is generally the dimensions of the components, which must not exceed the dimensions of the galvanizing kettle as well as the lifting capacity of the handling equipment in the galvanizing plant.

2.1 Continuous galvanizing

Some mass production items are suited to be galvanized in continuous systems [15]. Besides high productivity such lines offer the benefit of accurate setting of the process parameters. It allows you to efficiently check the basic parameters of the coating such as appearance, structure, and thickness. The best-known continuous galvanizing method is the Sendzimir process. In a process line (Figure 2.1), cold rolled steel sheet with a relatively clean surface is welded to form an endless strip and it is continuously degreased and pickled. Then it passes through a furnace where, in a reducing atmosphere, the material is subject to full annealing and oxides

Hot-Dip Galvanizing of Steel Structures. DOI: http://dx.doi.org/10.1016/B978-0-08-100753-2.00002-1

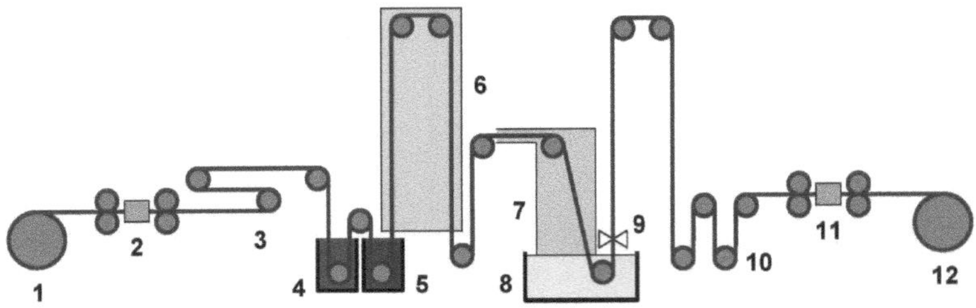

Figure 2.1 Diagram of a line for continuous sheet metal galvanizing. 1. strip unwinding, 2. welding, 3. leveling loop, 4. degreasing, 5. pickling, 6. full annealing, 7. surface activation in a reduction atmosphere, 8. zinc bath, 9. air wiping knives, 10. leveling, 11. shearing, 12. winding the galvanized strip into a coil.

are removed from its surface. In a protective atmosphere, the clean metallic strip is then immersed in a galvanizing bath and exits in a vertical orientation through mechanical air knives that wipe the coating surface. After leveling, the strip is wound up into coils.

Aluminum is generally added to the zinc bath in continuous galvanizing [19] as it takes priority in reacting with the substrate to form a thin, consistent, iron-aluminum alloy layer. That layer prevents further iron diffusion and its reaction with zinc suppresses the formation of brittle iron-zinc alloy phases so that the coating generally consists of relatively malleable pure zinc. This produces favorable mechanical properties for further forming of galvanized metal sheets. The thickness of the zinc coating on the surface of a continuously galvanized strip is approximately 10 to 35 μm.

2.2 Batch galvanizing

Significant portion of hot-dip galvanized steel parts is batch galvanized in commercial galvanizing plants. Batch hot-dip galvanizing is carried out in a standardized manner. In Europe, the technical conditions for batch hot-dip galvanizing are described by EN ISO 1461, which is a binding technical standard for the member states of CEN (Comité Européen de Normalisation).

2.2.1 Dry process

Most commercial galvanizing plants are equipped with a technology for suspension of batches of components by cranes and jigs (Figure 2.2) within the dry process (Figure 2.3). A benefit of this technology is relatively high productivity at a sufficiently high degree of mechanization. The dimensions of tanks of these plants are adapted to the market demand. If the design and production principles of hot-dip galvanized parts are observed, very large parts can be galvanized in these plants with transportation capacities becoming the main limiting factor (Figure 2.4).

Figure 2.2 Sling galvanizing.

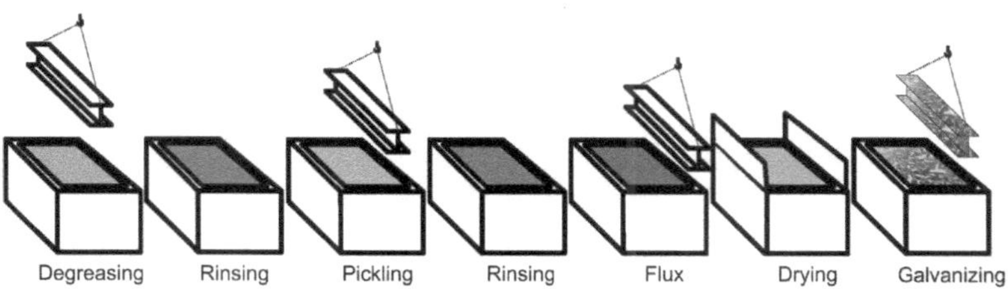

Figure 2.3 Schematic representation of the process sequence of operations in a dry galvanizing process line.

Figure 2.4 Hot-dip galvanized steel structure ready for shipment.

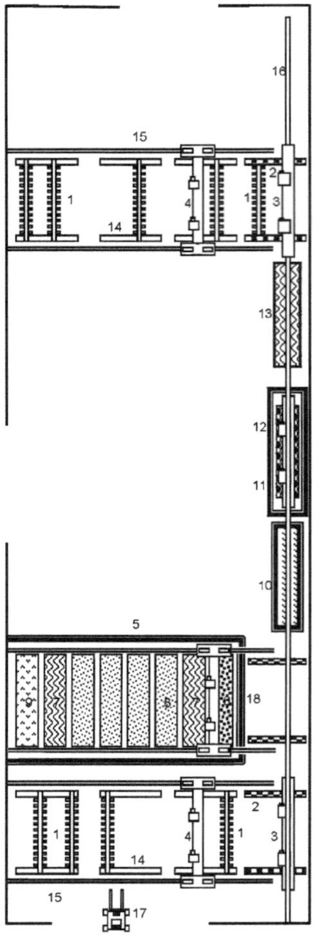

Figure 2.5 Linear arrangement of the process line of a hot-dip galvanizing plant.
1. suspension beam (for fixation of galvanized items), 2. transversal conveyor, 3. suspension crane, 4. overhead traveling crane, 5. enclosed chemical pre-treatment workplace, 6. degreasing bath, 7. rinsing, 8. pickling bath, 9. flux, 10. drying furnace, 11. enclosed galvanizing workplace, 12. galvanizing crane, 13. water cooling crane, 14. stand for putting the suspension beam aside, 15. crane track for the overhead crane, 16. single-beam crane track, 17. stacking truck, 18. high-speed roller door.

The hot-dip galvanizing line may be arranged linearly (Figure 2.5), U-shaped (Figure 2.6), or as a combination of these options. Prepared parts are first chemically pre-treated. Individual process tanks are arranged in accordance with the operation sequence from the degreasing bath through the pickling and rinsing baths to the flux bath. High-quality surface decontamination requires quite a lot of time; therefore the chemical pre-treatment workplace is always equipped with a higher number of pickling baths. In commercial galvanizing plants galvanizing is done in

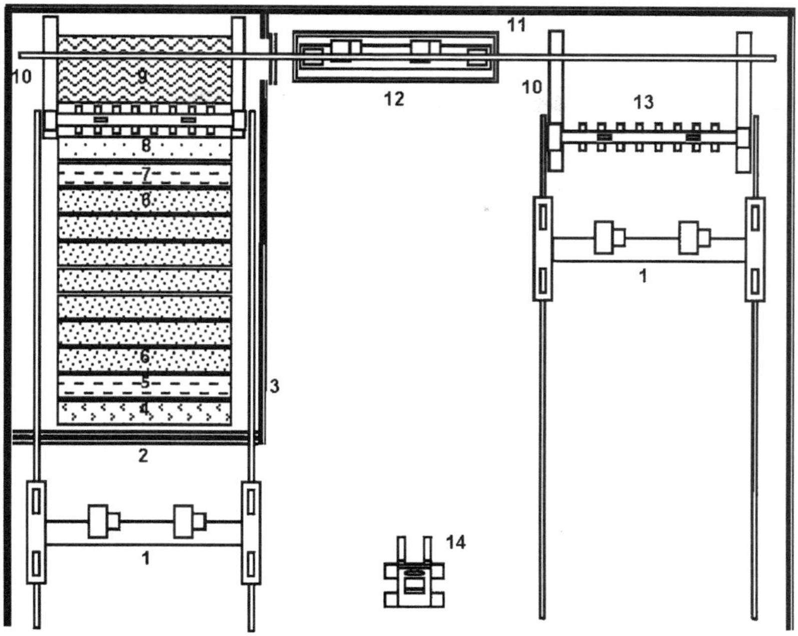

Figure 2.6 U-shaped arrangement of the process line of a hot-dip galvanizing plant.
1. overhead traveling crane, 2. high-speed roller door, 3. enclosed chemical pre-treatment workplace, 4. degreasing bath, 5. bath with rinsing water, 6. pickling bath, 7. bath with rinsing water, 8. flux, 9. drying furnace, 10. mechanical stand with a chain conveyor, 11. enclosure of the galvanizing workplace, 12. galvanizing workplace with a suspension crane, 13. suspension beam, 14. stacking truck.

batches that pass through the entire line as they were suspended on jigs at the beginning of the line. After the galvanizing bath, a cooling water bath may be used or the items are left to air-cooling.

2.2.2 Wet process

Some commercially galvanized production is galvanized in plants using the wet process technology (Figures 2.7 and 2.8) [15]. In spite of relatively low productivity of these plants and limitations of manual handling there are products whose galvanizing would not be possible in other equipment types due to their complex shape or lack of access to detailed openings. Unlike the dry process (see Section 2.2.1) the wet process line does not contain the drying operation; therefore this technology is called the wet process. Immediately after being removed from the pickling bath, in the wet state, the batch is immersed in molten zinc through flux covering a part of the zinc bath surface (Figure 2.9) and galvanized parts are removed from the zinc bath in the part separated by the slide (Figure 2.10).

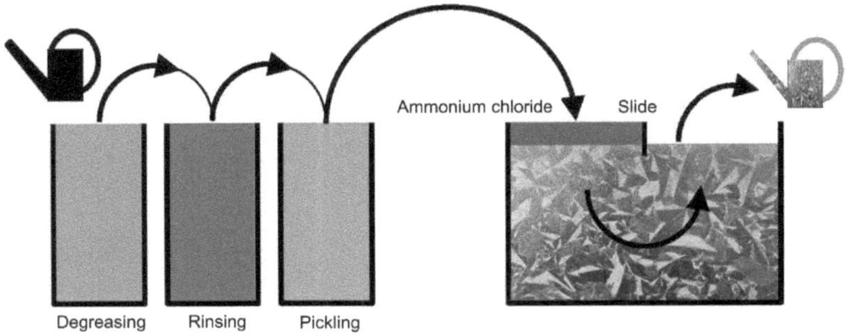

Figure 2.7 Operation sequence diagram of the wet galvanizing process.

Figure 2.8 Modern wet process galvanizing workplace in the ALKA Chotěboř galvanizing plant.

Figure 2.9 The space for immersion of parts to be galvanized in the zinc bath through a sal-ammoniac layer is equipped with a powerful extraction device with an efficient dust filter.

Figure 2.10 Galvanized parts are removed in the bath part separated by the slide through clean surface.

2.2.3 Centrifuge (spun) galvanizing

Hot-dip centrifuge galvanizing is used for small parts [15]. The batch is first subjected to the usual chemical pre-treatment, generally in perforated plastic containers that are resistant to chemicals. Then, after dripping and drying they are transferred into cylindrical galvanizing baskets that are immersed in the zinc bath. After removal, the cylindrical basket is quickly transferred to a centrifuge where excessive zinc is centrifuged off the surface of the galvanized parts (diagram; see Figure 2.11). In modern equipment the basket is suspended on a rotary drive unit and the centrifuging is done immediately after removal from the bath over the galvanizing bath itself (Figure 2.13).

Centrifuge galvanizing may be carried out as a low-temperature process at the zinc melt temperature up to 480°C or as a high-temperature process [20], when the zinc bath temperature must be above 530°C (usually 550°C). After centrifuging the batch is cooled in water (Figure 2.14). Quick cooling of the batch during high-temperature galvanizing is necessary to avoid further dwell of the batch at the peritectic transformation temperature when the mixture of phase δ and liquidus (molten zinc) produces phase ζ (see also Sections 4.3 and 5.4.6).

An advantage of pickling in plastic containers (Figure 2.12) is elimination of the stripping operation of galvanizing baskets, which do not get in contact with the pickling solution and thus have a longer service life.

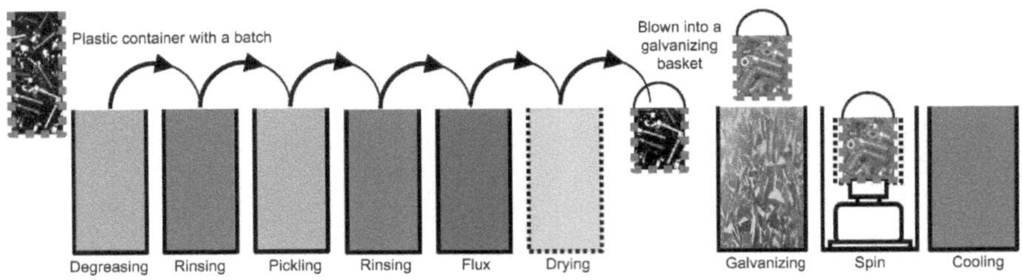

Figure 2.11 Operation sequence diagram in a technological line for hot-dip centrifuge galvanizing.

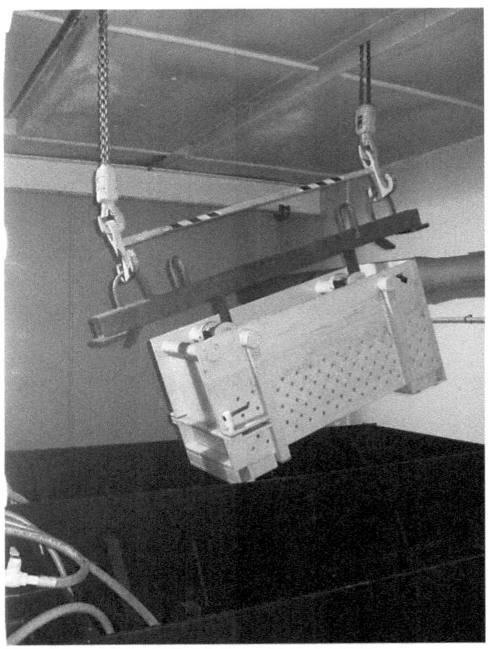

Figure 2.12 Chemical pre-treatment of small parts in a plastic container before centrifuge galvanizing.

The reason for high-temperature galvanizing at a temperature above 530°C is that at these temperatures phase ζ is not formed (see Section 4.3) and a consistent layer of phase δ prevails in the coating, which may be overlaid by a layer of loosened fine crystals of this phase interfused with pure zinc after longer immersion periods (see Section 5.4.6). The coating has a relatively low thickness range and very good mechanical resistance. Therefore, high-temperature centrifuge galvanizing is used for coating of parts with a mating surface, especially screws (Figure 2.15). Before galvanizing the thread is produced with limit tolerances that take into account the known thickness of the applied coating and the threaded

Figure 2.13 The parts may already be centrifuged during the removal of the basket with the batch from the zinc bath.

Figure 2.14 Cooling of centrifugally galvanized parts in water.

connection is functional. Since pure zinc is centrifuged off the coating, high-temperature galvanized parts are characterized with a matte, dull shade of the coating consisting of phase δ, which exhibits favorable mechanical properties for application on fasteners. The thickness of coating applied at high temperatures is relatively low, approximately 35 to 45 μm. To maintain balanced anti-corrosion resistance of the entire structure, centrifuge galvanized fasteners should be coated with organic paint after the assembly. Generally, a touch-up coating material with a high content of zinc dust is applied for this purpose. This material is supplied for possible repair of coating defects.

Figure 2.15 Centrifugally galvanized fasteners.

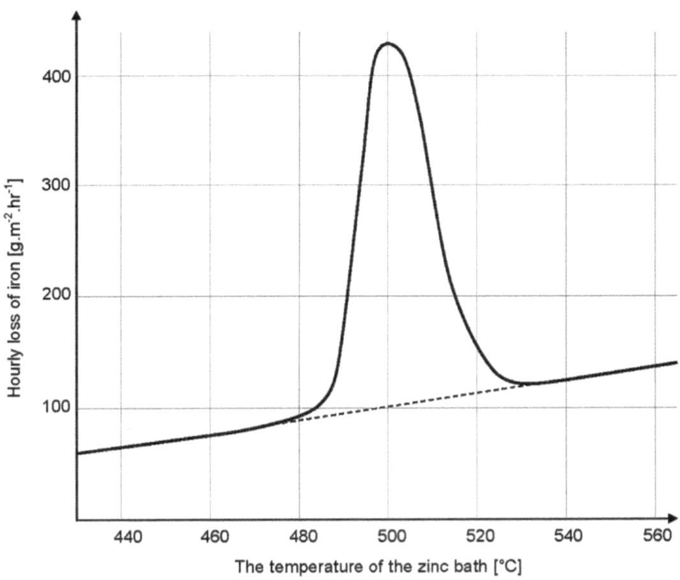

Figure 2.16 Iron losses due to its dissolution in a zinc bath depending on the temperature.

The temperature range between 480 and 530°C is unsuitable for hot-dip galvanizing because in this temperature interval the rate of dissolution of iron in molten zinc abruptly rises (Figure 2.16). This leads to a material change of geometry of the galvanized parts accompanied by simultaneous deterioration of zinc in the zinc bath. It combines with dissolved iron in the ratio of 13:1, forming hard zinc (phase ζ), which is a waste product.

Chemical pre-treatment

3

The metallurgical reaction leading to the formation of the galvanized coating can only be successful if the surface of the parts is clean and without any contaminants and suitably activated. In modern galvanizing plants, chemical pre-treatment (Figure 3.1) is carried out in an enclosed workplace. This guarantees reliable collection of any emissions released during pickling. Thus, negative environmental influences are minimized and at the same time favorable working conditions for galvanizing plant operators are provided. The basic chemical pre-treatment operations are degreasing, pickling, and (in the case of the dry process) application of flux. Individual steps are followed by rinsing to avoid transfer of chemicals between the process baths, which would lead to their deterioration.

The process bath tanks of the chemical pre-treatment must be resistant to extreme corrosion stress. Merely lining the surface of a steel tank with an anti-corrosion acid-resistant material is insufficient for its required service life. In modern galvanizing plants, the jackets of the process tanks are made of welded polypropylene, which has a high chemical resistance, and the tanks are reinforced with steel frames.

3.1 Degreasing

The first operation in the chemical pre-treatment workplace is degreasing [21]. First, grease is removed from the surface of the parts in a degreasing bath (Figure 3.2) as grease would hinder perfect wetting in the pickling bath. Commercial galvanizing plants use alkaline or acidic degreasing agents.

Alkaline degreasing is done in an aqueous solution of an alkaline degreasing agent, usually with an admixture of surfactants. The main components of the alkaline agents are sodium hydroxide and potassium phosphate. The degreasing action consists of dissolution of grease and its emulsification. To achieve the required degreasing efficiency, the temperature of the degreasing bath must be maintained above 50°C. Therefore, this degreasing method has a high energy demand. During its useful life the bath gradually fills with emulsified grease and when a certain level of saturation has been achieved (approximately 0.5%), it ceases to work. The saturated solution must be replaced with a new bath. The useful life of the degreasing bath can, to a certain extent, be extended by installation of a mechanical device for skimming grease from the surface. Alkaline degreasing must always be followed by a rinse to prevent deterioration of the acidic pickling bath.

Acidic degreasing is becoming more and more popular. Commonly used agents that also utilize surfactants usually make use of the action of phosphoric acid in a mixture with nitric, sulfuric, or hydrochloric acid. Advantages of acidic degreasing

Hot-Dip Galvanizing of Steel Structures. DOI: http://dx.doi.org/10.1016/B978-0-08-100753-2.00003-3

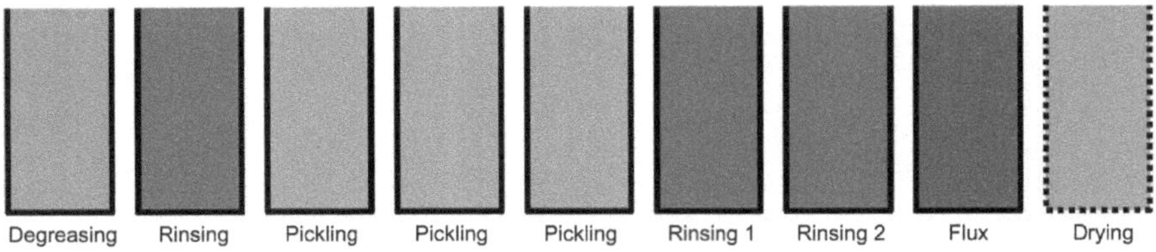

Figure 3.1 Operation sequence diagram of the chemical pre-treatment process before hot-dip galvanizing.

Figure 3.2 Excessive greasing of a part cannot be removed using the standard degreasing process.

baths are that the pickling process already starts during the degreasing and, more importantly, that grease does not dissolve in the bath but precipitates and settles on the tank bottom in the form of sediment. From that point the sediment is removed at regular intervals when the bath has been pumped out and the original bath with a virtually unlimited service life is returned to the tank. Only the loss caused by removal of the degreased items is replenished. Acidic degreasing agents work at the ambient temperature; thus they do not require costly heating and reduce the production of hazardous waste. However, they may have lower efficiency than alkaline agents.

3.2 Pickling

A perfectly clean metallic surface of a degreased part is achieved by subsequent pickling in an inorganic acid. The most frequently used pickling agent in commercial galvanizing plants is hydrochloric acid in a concentration of up to 14%. To increase efficiency, the pickling bath is formulated with a certain proportion of ferrous chloride, which ensures better dissolution of scale covering the surface of parts to be galvanized. In the course of its useful life the content of iron in the pickling bath increases while the content of free HCl decreases. To achieve optimum pickling time the contents of both the components in the solution must be maintained in a suitable mutual ratio. Generally, based on an analysis of the chemical composition the content of free HCl is increased in such a way that after extraction of a predetermined volume of the pickling solution the same quantity of concentrated acid is added to the bath. Inhibitors are added to the pickling bath that limit dissolution of iron during pickling and reduce the rate of saturation of steel with hydrogen.

In some plants, sulfuric acid is exceptionally used for pickling. To achieve a sufficient pickling effect it must be heated up to a temperature above 50°C, which means increased energy demands and production of emissions.

3.2.1 Principles of operating pickling baths based on hydrochloric acid

The purpose of pickling is to achieve a perfectly and chemically clean metallic surface of the steel to facilitate the metallurgical reaction that produces the alloy coating. The surface of the steel part is usually covered with scale and corrosion products (Figure 3.3), which can be efficiently removed by pickling, usually in hydrochloric acid. During this process the iron oxides become gradually separated from the base material.

Scale contains three basic layers of iron oxides as explained in Table 3.1.

The shares of individual iron oxides may be different in the scale layer. To efficiently remove wüstite, which is always closest to the substrate, the pickling solution must get access to it (Figure 3.4). Scale predominantly consisting of magnetite (Figure 3.5) requires very long pickling times. Parts on the surface where wüstite predominates, which dissolves readily in the pickling solution (Figure 3.6), need

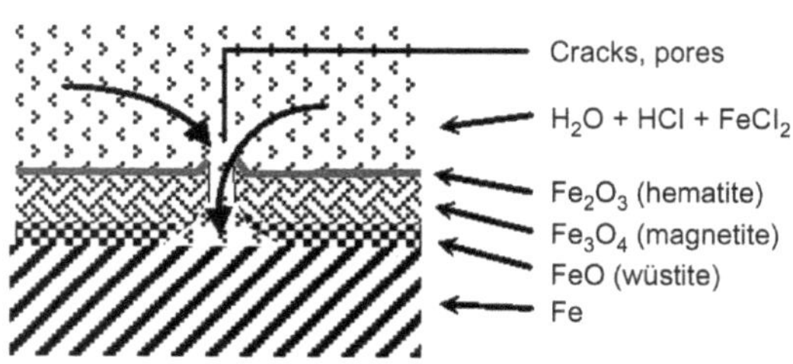

Figure 3.3 Principle of scale removal by a pickling solution.

Table 3.1 **Scale layer structure on steel**

Substance	substrate	wüstite	magnetite	hematite
Composition	Fe	FeO	$FeO + Fe_2O_3$	Fe_2O_3
Solubility	difficult	very good	difficult	very difficult

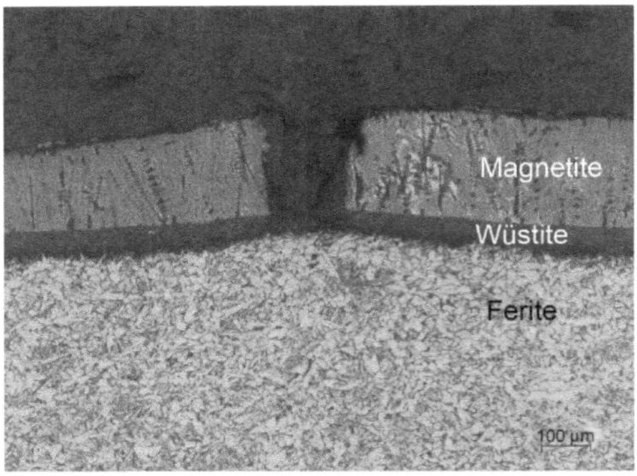

Figure 3.4 Scale layer with a crack.

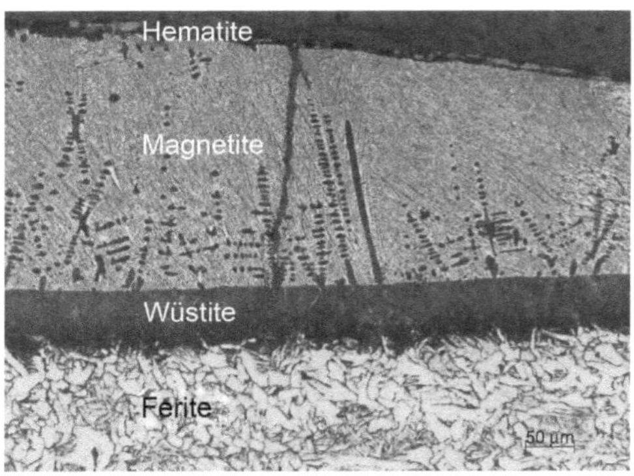

Figure 3.5 Scale with a high proportion of magnetite.

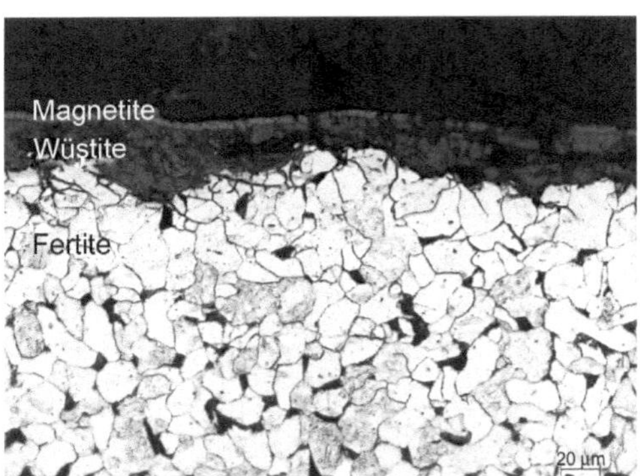

Figure 3.6 Scale with a high proportion of wüstite.

relatively short pickling times to achieve a clean metallic surface. In modern metallurgical plants, scale is removed from the surface of rolled products with the use of efficient methods and the occurrence of strongly scaled metal sheets is relatively rare.

The wüstite layer of the scale dissolves first, as ferrous oxide is most readily soluble in the pickling solution:

$$Fe(OH)_2 + 2HCl \rightarrow 2H_2O + FeCl_2$$
$$FeO + 2HCl \rightarrow H_2O + FeCl_2$$

During pickling, reaction of pure iron with the hydrochloric acid cannot be excluded, which produces hydrogen. This might produce harmful effects in the steel and any coating applied onto it (see Sections 3.6 and 10.1):

$$Fe + 2HCl \rightarrow FeCl_2 + H_2$$

Therefore, inhibitors are often added to pickling baths to inhibit the reaction of iron with hydrogen chloride and so reduce the rate of saturation of the substrate with hydrogen.

There are basic requirements to pickling related to environment and production process. The basic pickling bath management principles must comply with these requirements:

a. high pickling efficiency (the shortest possible pickling time);
b. high acid utilization rate;
c. minimize negative environmental influences;
d. trouble-free and price efficient disposal of used acid.

In the course of the useful life of the pickling solution, the content of free acid in the solution decreases while the content of ferrous chloride increases. The initial concentration of the acid in the pickling solution should not exceed 150 g/l for environmental reasons. Figure 3.7 shows the position of the boundary line for the concentration of HCl emissions amounting to 10 mg/m^3 at various temperatures of the pickling solution, depending on the free HCl content [22]. A higher temperature of the pickling bath increases pickling efficiency. If pickling baths are to be operated at higher temperatures (above the boundary line), efficient extraction and separation of HCl emissions must be ensured so that the limit concentration of air pollutants in the released air is not exceeded.

During the dissolution of rust and scale in hydrochloric acid, considerable quantities of ferrous chloride ($FeCl_2$) are formed in the pickling bath, which produces a green to dark green color in the used bath. During the pickling process the insoluble brown ferric chloride $FeCl_3$ is initially formed, which is, however, gradually reduced to the green $FeCl_2$. What is important is that as the free HCl content in the pickling solution goes down, the solubility of $FeCl_2$ increases. However, ferrous chloride only dissolves in hydrochloric acid to a limited extent. As soon as the pickling solution is completely saturated with ferrous chloride, further pickling under unchanged conditions (without increasing the bath temperature) is no longer

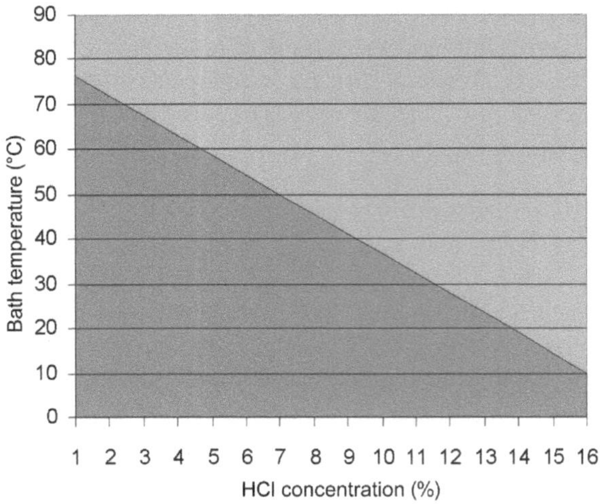

Figure 3.7 Limit line for the concentration of HCl emissions of 10 mg/m^3.

possible and a change of the pickling bath composition does not lead to any improvement. A pickling solution saturated with ferrous chloride (containing approximately 2% of remaining HCl) is sent to be disposed of in an environment-friendly way. Figure 3.8 illustrates the dependence of the efficiency of the pickling solution on its chemical composition. To keep the pickling efficiency (rate) as high as possible, the respective decrease of the free acid content must correspond to the increasing content of iron.

3.2.2 The "cross rule"

For efficient pickling bath management the solubility dependences in the ternary $H_2O + HCl + FeCl_2$ system must be understood. To ensure sufficient efficiency of a new pickling bath, it must be formulated with a sufficient content of $FeCl_2$ (approximately 50 g/l). Based on an instantaneous analysis of the chemical composition of the pickling bath it is made stronger or diluted to optimize HCl and Fe proportions. Individual quantities that should be mixed to achieve a certain concentration can be determined with the use of the "cross rule" of mixing in accordance with the diagram in Figure 3.9.

To obtain a C% concentration from the A% and B% solution you must mix (A−C) parts of the B% and (C−B) parts of the A% solution.

The parts can be expressed in mass if the concentration is given in weight percent or expressed in volume if the concentration is given in volume percent.

The content of zinc should be kept to a minimum in pickling baths as its presence decelerates the pickling process. Therefore, de-zinced (stripped) jigs/hooks should be used to hang parts to be galvanized. Removal of zinc coating from jigs/hooks and rejected galvanized parts should be done in a separate pickling bath. The

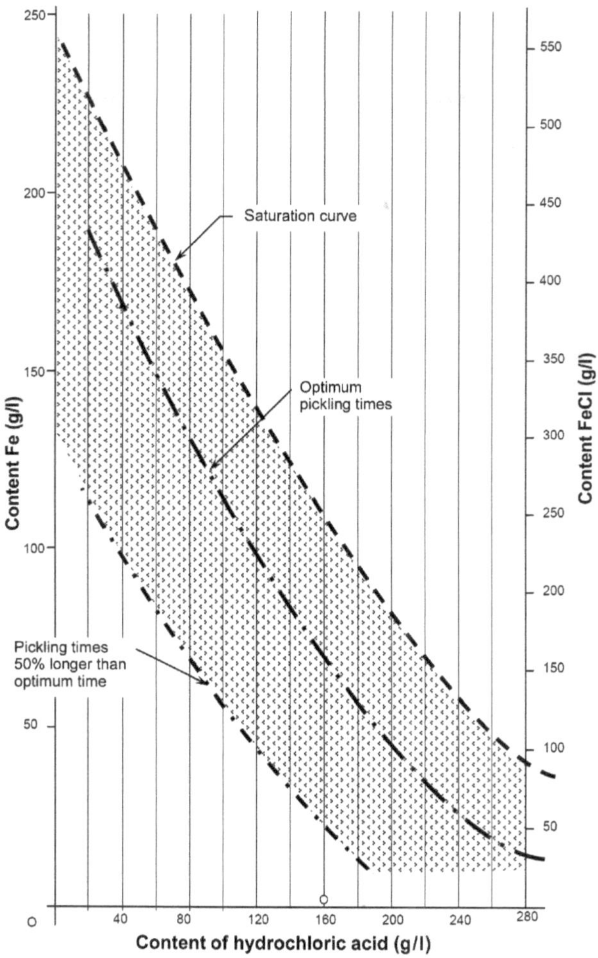

Figure 3.8 Optimization of pickling bath composition.

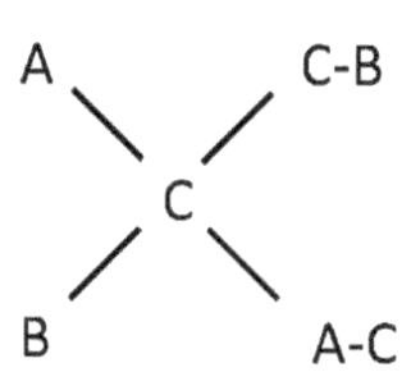

Figure 3.9 The "cross rule."

environmental aspect is also important in this respect as saturated mixed solutions are more difficult to dispose of. Zinc chloride produced by proper operation of a separated stripping (de-zincing) bath is an important raw material from which zinc can be efficiently recycled for reuse in the chemical or pharmaceutical industry. To extract zinc from the stripping bath efficiently the iron content in the bath must be kept to the minimum.

3.3 Rinsing

Within the dry process the pickling operation is always followed by a rinse to prevent transfer of ferrous chloride from the parts to the flux and subsequently to the zinc bath, where is has negative effects as it promotes increased formation of dross (phase ζ). Some modern galvanizing plants have a two-stage rinsing station where every batch is immersed in the first and then in the second rinsing bath after dipping in the pickling solution. During operation, water from the first rinsing bath is used to replenish water evaporated from the pickling bath and the withdrawn quantity is then replenished from the second rinsing bath where clean service water is replenished after that. This is to make sure that the plant does not produce any wastewater that would otherwise have to be treated in an expensive but environment-friendly way.

3.4 Flux

In the typical dry process, before parts to be galvanized are immersed in molten zinc, flux is applied to their surface by their immersion in an aqueous solution of zinc chloride and ammonium chloride, which are contained in the bath in the ratio of approximately 3:2. The salt content in water should be approximately 400 g/l [23]. The temperature of the flux is generally maintained under 50°C. The applied flux is dried and a protective film is created on the surface that prevents further oxidation.

During immersion of parts into the molten zinc at the temperature of approximately 450°C, ammonium chloride (NH_4Cl) decomposes to ammonium and ammonia (NH_3), which escapes, and to hydrogen chloride (HCl), which mainly dissolves zinc and aluminum oxides on the surface of the zinc bath enabling direct contact of the steel surface with molten zinc. The task of zinc chloride ($ZnCl_2$) is to reduce the surface tension of zinc, activate the surface, and initiate the metallurgical reaction between iron and zinc that occurs during the immersion of the batch in the zinc bath and produces the iron-zinc alloy coating.

Iron contained in flux has an unfavorable impact on application of the coating in the zinc bath as it is the cause of increased production of dross (a recyclable waste product of galvanizing), which is formed from the zinc contained in the galvanizing bath. The flux has a slightly acidic reaction; its pH is approximately 4.5 and it means that during a longer immersion period of the batch in the flux bath, any iron contained in the flux dissolves. On the other hand, transfer of ferrous chloride to the flux from the pickling solution cannot be reliably excluded either. Here, unsuitable design of galvanized parts characterized by pockets and corners with unsuitable or missing drainage plays a very negative role. The iron content in the flux must be kept at very low values. As a suitable method of flux regeneration, conversion of bivalent iron from ferrous chloride to trivalent iron by means of hydrogen peroxide H_2O_2 has proved successful. Within the process, the pH of the solution must be reduced to pH2 to initiate hydrolysis of iron producing the insoluble ferric hydroxide $Fe(OH)_3$, which can be then collected by a filter press. Then, the pH must be adjusted to the required value of pH4.5, e.g., by adding ammonium hydroxide (NH_4OH) or zinc oxide (ZnO). A similar effect to that of hydrogen peroxide can be achieved with the use of potassium permanganate. This method produces the insoluble manganese dioxide, which is filtered off together with ferric hydroxide [24].

3.5 Drying

In the dry process, flux applied on parts to be galvanized must be dried to prevent splashing of zinc during immersion of the batch into the galvanizing bath. Drying is most frequently done in a booth with efficient ventilation, and waste heat from the heating of the galvanizing bath is preferably used for heating. Immediately after drying, the items are immersed into the molten zinc.

Note

Due to the elevated temperature of the gases and high pollution level of the air in the drying booth the material that the drying booth is constructed from is exposed to extreme corrosive loading.

3.6 Harmful effects of hydrogen

During the reaction of iron with the hydrogen chloride contained in the pickling solution nascent (freshly generated) hydrogen is produced, which passes into an area that has a lower value of its partial pressure [25]. Such an area may also be steel, where hydrogen gradually penetrates during the pickling process and forms an interstitial solution in it. A part of the hydrogen inside the steel remains in the atomic form and after completion of the pickling it escapes from the substrate again. However, some atoms combine to form molecules in places of lattice imperfections and vacancies, inclusions, etc. (hydrogen recombines), and significantly increase their volume (see also Section 10.1).

3.6.1 Blisters and foam

Blistering [26] of the zinc coating (see Section 7.8.6) is caused by hydrogen escaping from steel during the galvanizing process. Given its very small dimensions, atomic hydrogen penetrates into various materials very easily. From places with a higher hydrogen partial pressure it passes to places where its partial pressure is lower. Therefore, during pickling hydrogen penetrates into steel where its quantity is relatively lower than in the pickling solution and later, especially when the steel is heated in the zinc melt, escapes from the steel again. At the same time, the layer of the originating alloy phase ζ in the coating represents an area of frequent imperfections in the crystalline structure. In such places atomic hydrogen quickly recombines. As the hydrogen emerges, when the layer of zinc adhering to the surface of the part solidifies, a barrier is formed against escaping of recombined hydrogen and blisters can appear in the coating (see Section 7.8.6).

The cause of foam formation (see Section 6.1.8) is also hydrogen escaping from the substrate, which is in this case blocked by a mushy substance formed on steel containing silicon in the Sandelin area (see also Section 6.1.1).

3.6.2 Defects of organic coatings

Atomic hydrogen, which penetrated into steel during pickling, only escapes from it partly during the immersion in the zinc bath. A considerable amount is released from steel later – a few days after the galvanizing. Every next heating of the substrate accelerates this process. This may cause problems, e.g., during baking of powder coatings applied on the galvanized coating. Baking of an organic material causes its sintering and this barrier blocks free escape of hydrogen, which recombines in the porous zinc coating and due to increasing of its volume causes formation of bubbles and craters in the organic material layer (Figure 3.10) [27].

Figure 3.10 Defects in plastic powder coating caused by hydrogen.

3.6.3 Hydrogen embrittlement and hydrogen corrosion

Hydrogen that recombined in substrate defects achieves a pressure of up to several hundreds of megapascals [28, 29]. During the hot-dip galvanizing process this pressure rises even further and the yield point of the material is locally exceeded. A higher risk of hydrogen embrittlement is present in steels with a higher yield point, which put up increased resistance to possible plastic deformations, show a difficult relaxation capability, and, in extreme cases, a brittle fracture may occur (see also Section 10.1).

Hydrogen that combines with free carbon precipitated at grain boundaries, producing methane, causes "hydrogen corrosion." Methane is not capable of further diffusion and accumulates in the place of its generation where it causes strong stress. Welds and heat-affected weld areas exhibit an increased risk of hydrogen corrosion.

Hot-dip galvanized coating formation

4

An intermetallic coating applied at elevated temperature by immersion in molten metal is generated on the principle of a metallurgical reaction between the substrate and the applied metal.

During hot-dip galvanizing, the iron-zinc alloy coating grows on the galvanized parts and is the result of a relatively complex process of mutual diffusion of atoms of both the metals. This produces elementary intermetallic bonds and subsequent phase transformations. These processes occur in the surface layer of the metal to be galvanized on the one hand, further at the boundary between the substrate and molten metal as well as in the zinc melt surrounding the galvanized part in its immediate vicinity. When the part is immersed in the zinc melt, zinc and iron react, producing intermetallic iron-zinc alloy phases. In batch hot-dip galvanizing, this metallurgical reaction takes place at a temperature of approximately 450°C.

4.1 Iron–zinc system

The most frequently hot-dip galvanized material is ferritic or ferritic-pearlitic structural steel in various quality grades. Iron, like zinc, is a transition metal. The binary iron–zinc system (Figure 4.1) is characterized by the partial solubility of one component and virtually complete insolubility of the other component. While ferrite forms a solid substitution primary solution α with zinc, this may, at the peritectic transformation temperature (782°C) contain up to 41% of zinc; on the other hand, the maximum content of iron in solid zinc is 0.03% at the most. The iron–zinc system is a system with three peritectic transformations and one peritectoid transformation. It is characterized by a considerable difference of the melting point between the two components. The temperature at which the alloy coating is formed is above the melting temperature of zinc (419.4°C) but below the melting point of iron (1538°C). Thus, within batch hot-dip galvanizing at temperatures around 450°C, equilibrium conditions for a peritectoid transformation or any of the peritectic transformations cannot be achieved. The conditions for a peritectic transformation (a liquid and solid phase are transformed to a new solid phase under equilibrium conditions) or for a peritectoid transformation (two solid phases are transformed to a new solid phase under equilibrium conditions) can only occur during so-called "high-temperature" galvanizing. During conventional galvanizing, zinc and iron can only be mutually bonded to intermetallic alloys on the basis of diffusion processes.

Hot-Dip Galvanizing of Steel Structures. DOI: http://dx.doi.org/10.1016/B978-0-08-100753-2.00004-5

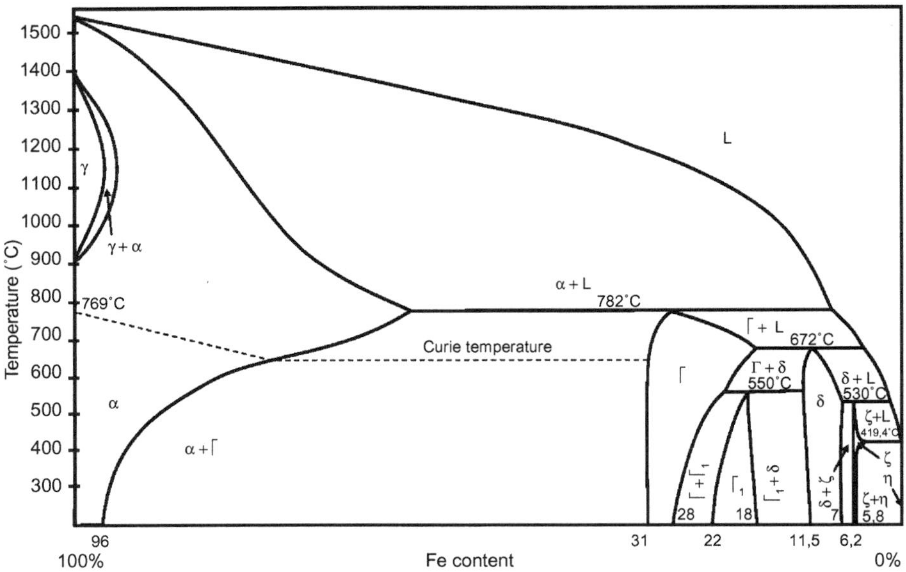

Figure 4.1 Iron–Zinc binary diagram.

The Fe–Zn binary system has only become the object of more detailed research studies quite recently. The goal of this research is to use modern, very accurate analytic methods to verify some phase transformation theories. Research conducted by B. P. Burton, P. Perrot, and H. Okamoto [30] represents a great contribution to our knowledge of the Fe–Zn system. X. Su, N-Y Tang, and J. M. Toguri later developed their findings, publishing their results in their work "Thermodynamic evaluation of the Fe–Zn system," issued by Elsevier in 2001 [31]. According to their diagram, the phase boundaries correspond to the observation in Mösbauer spectra during annealing of galvanized samples performed by M. Zmrzlý, O. Schneeweiss, V. Houbaert, and J. Fiala in 2005 [32]. Also, J. Nakano, in his work "Thermodynamics of the Zn-Fe-Al-X system," published in 2005 [33], adopts the Okamota mode for the binary iron–zinc system and confirms the delimited stability boundaries of individual phases. The conclusions of the research [30, 31] were later analytically verified by P. Rajak, U. Tewary, S. Das, B. Bhattacharya, and N. Chakraborti, who described the results of their analyses and experiments in their article "Phases in Zn-coated Fe analyzed through an evolutionary meta-model and multi-objective Genetic Algorithms," published by Elsevier in 2011 [34]. This work confirms the correctness of the results achieved by X. Su, N-Y Tang, and J. M. Toguri.

Iron and zinc form two-component alloy phases Γ, Γ_1, δ, and ζ (Figure 4.2), [31–35], (see Table 4.1). Both phases Γ and Γ_1 crystallize in a cubic, body-centered system and differ in the size of the lattice constant. The lattice constant of phase Γ_1 is twice that of phase Γ [34].

During the immersion time of steel in the zinc bath, which is approximately 5 to 10 minutes depending on the batch's characteristics, two coating layers are formed

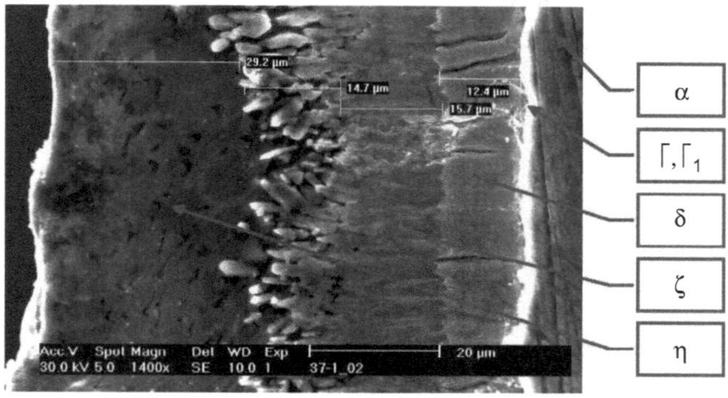

Figure 4.2 Basic composition of the coating applied on non-silicon-killed steel.

on its surface after sufficient heating – closer to the substrate there is the alloy phase δ and phase ζ is formed on it. At approximately 450°C, which is the usual temperature of the zinc bath, the solubility of zinc in ferrite significantly increases and, depending on the structure and chemical composition of the steel, a more or less saturated solid solution of zinc in iron α is formed in its surface layer.

Phases Γ and Γ_1, produced by direct synthesis due to diffusion, are often imperceptible in coatings applied by batch galvanizing. If the ferrite has been sufficiently saturated with zinc and, if after the removal from the zinc bath the galvanized part is cooled down slowly, conditions may be created for segregation of the phase Γ from the oversaturated solid solution of zinc in iron α (see Section 5.1). It is the electron compound Fe_3Zn_{10} [34], which may be generated at a suitable ratio of the valence electrons to the number of atoms. Phase Γ, the same as ferrite (iron α), crystallizes in the cubic crystalline system with a body-centered lattice and exhibits high adhesion to the substrate.

Figure 4.3 shows, at 20,000 times magnification, the results of an analysis performed to prove the presence of phases Γ and Γ_1 in coating applied on non-silicon-killed S235 steel with an immersion time in the zinc bath of 8 minutes. The detail, showing distinct islands of ferrite and shrinkage porosities in the substrate in the place of the phase boundary between the substrate and coating, illustrates well the mechanism of diffusion dissolution of steel in a liquid metal, causing the occurrence of Kirkendall vacancies (see also Section 7.8.2). Figure 4.4 is a metallographic image of hot-dip galvanized coating in an area adjacent to the substrate on a sample of non-silicon-killed S235 steel exposed to the action of liquid metal for 120 hours.

Figure 4.5 shows the boundary between the substrate and the coating. A zinc-saturated surface layer of steel is a prerequisite for segregation of phase Γ during sufficiently slow cooling. The spongy structure of the steel surface resembling a selective corrosive attack of a galvanized metal is related to the occurrence of "Kirkendall vacancies" that are generated by the prevailing diffusion of iron atoms. The flow of iron atoms is compensated for by the flow of vacancies into steel (see also Section 7.8.2).

Table 4.1 **List of iron-zinc alloy phases**

Phase	α	Γ	Γ_1	δ	ζ	η
Composition	Fe	Fe_3Zn_{10}	Fe_5Zn_{21}	$FeZn_{10}$	$FeZn_{13}$	Zn
% Fe	59 ÷ 100	19 ÷ 31	18 ÷ 22	7 ÷ 11	5.8 ÷ 6.2	—
Crystalline structure	cubic (body-centered)	cubic (body-centered)	cubic (body-centered)	hexagonal (columnar)	monoclinic	hexagonal (close packed)
Mechanical characteristics	ferritic steel	strongly adhesive	strongly adhesive	consistent	hard, brittle	tough

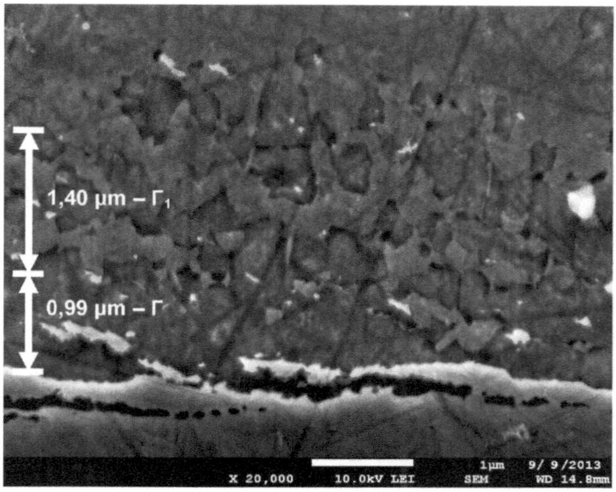

Figure 4.3 The presence of phases Γ and Γ_1 in the coating can be generally be proven only with the use of very precise analytic methods.

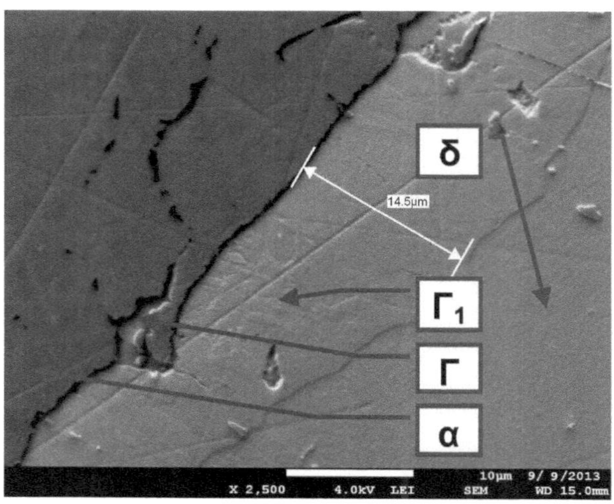

Figure 4.4 Alloy phases Γ and Γ_1 in a steel sample exposed in zinc bath for 120 h.

Note

The coating in Figure 4.5 was applied on a sheet of S355 steel with the thickness of 10 mm and with the steel composition C 0.185%, Si 0.385%, Mn 1.379%, P 0.018%, S 0.017%, Al 0.036% at the bath temperature of 445°C and the immersion time in the bath of 5 min.

Figure 4.5 Detail of the phase boundary between the substrate and coating in S355 steel.

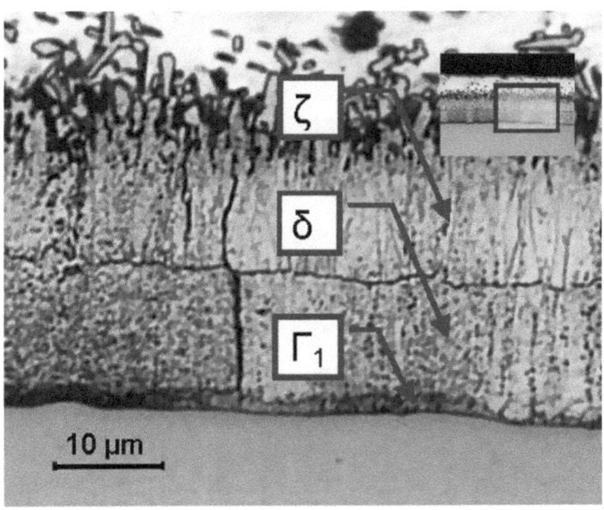

Figure 4.6 Detail of phase Γ_1 on S235 steel.

In phase δ iron and zinc may mutually bond in various proportions. Closer to the substrate the iron content is higher, and at the same time, at higher temperatures the iron in the alloy dissolves in higher concentrations than at ambient temperature. During slow cooling from the galvanizing temperature, a part of the δ-phase layer saturated with iron is transformed to phase Γ_1, which is separated from phase δ (and from the substrate) by a pronounced phase boundary in a metallographic sample (Figure 4.6).

Note

The coating in Figure 4.6 was applied on a sheet of S235 steel with the thickness of 20 mm and with the steel composition C 0.096%, Si 0.012%, Mn 0.882%, P 0.008%, S 0.009%, Al 0.028% at the bath temperature of 445°C and the immersion time in the bath of 5 min.

Metallography shows more frequent presence of phase Γ_1 in the coating, which, like phase Γ, crystallizes in the cubic, body-centered system but differs from phase Γ in that its lattice constant is double in size. Similarly to phase Γ, phase Γ_1 (Fe_5Zn_{21}) [34] can be generated by direct synthesis of zinc with iron diffusing from the substrate. Phase Γ_1 can also be generated by transformation from the alloy phase Γ during its slow cooling and/or segregation from phase δ. Figure 4.6 shows a clear layer of phase Γ_1 with phase Γ present next to it without a perceivable phase boundary in the area adjacent to the substrate. Its presence can only be proven with accurate analytic methods.

The composition of the alloy phase δ is $FeZn_{10}$ [34]. It is characterized by close packed hexagonal crystals in the form of prisms adjacent to each other (in a honeycomb-like way) that are able to form a consistent layer with favorable mechanical properties. Coating with significant presence of the δ phase layer offers good resistance to flaking caused by reckless handling.

Another alloy layer in the coating is phase ζ with a $FeZn_{13}$ [34] composition, which crystallizes in the monoclinic crystalline system. Unlike other iron-zinc alloy phases, phase ζ shows a very high homogeneity at ambient temperatures; its iron content fluctuates in a very narrow range (approximately 0.5%). In coatings on **non-silicon-killed steels** the crystals of phase ζ have the form of relatively fine needles growing with a perceivable phase boundary on the δ phase layer. With their arrangement they resemble a lawn that is overgrown with pure zinc on its surface. On the other hand, on **silicon-killed steels** phase ζ has a predominant share in the coating (the presence of phase δ in the coating of these steels is insignificant). On silicon-killed steels phase ζ consists of disorderly arranged, relatively compact monoclinic crystals. As a very thin layer of phase δ or disintegrated crystals of phase ζ do not represent a sufficiently efficient barrier to iron diffusion, the growth of the ζ phase layer is not decelerated on these types of steel and after longer immersion times the coating thickness gets bigger. Between individual monoclinic crystals of phase ζ there are numerous vacancies that reduce coherence of the coating and may be the cause of its reduced adhesion to the substrate.

4.2 Conventional (low-temperature) galvanizing

On non-silicon-killed steels, phase δ grows in a consistent, relatively thick layer that separates the substrate from the zinc melt and considerably decelerates diffusion of atoms of both the metals. On the other hand, on silicon-killed steels the layer of phase δ in the coating does not significantly develop during the immersion time in the zinc bath; it remains relatively thin and often inconsistent.

On non-silicon-killed steels the layer of phase ζ does not grow during the immersion time in the bath and its development is soon decelerated. Crystals of phase ζ are relatively fine. In the vicinity of phase δ they are closely packed and toward the surface they separate to the needle-like form. On silicon-killed steels the relatively thin and often inconsistent layer of phase δ does not prevent continuing diffusion of iron during the coating application and the thickness of the alloy ζ phase layer ($FeZn_{13}$) grows with time with an approximately linear dependence.

Figures 4.7–4.10 compare gradual coating formation on samples of steel: a) non-silicon-killed and b) silicon-killed. While on sample a) the phase δ layer gradually grows up to the thickness when the barrier created by it stops the

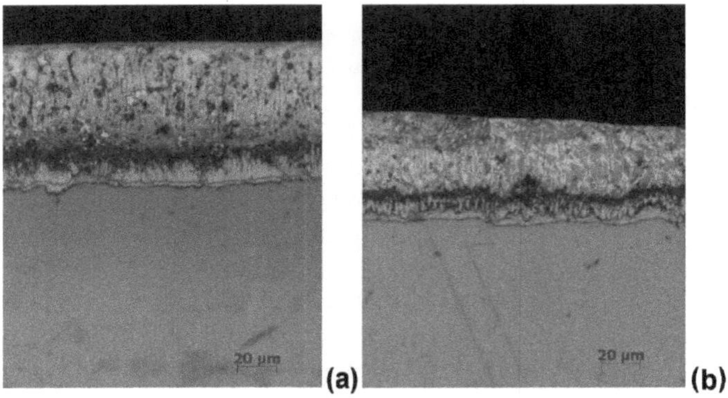

Figure 4.7 Coating after immersion time of 30 s: a) S235 steel (Si 0.012%), b) S355 steel (Si 0.196%).

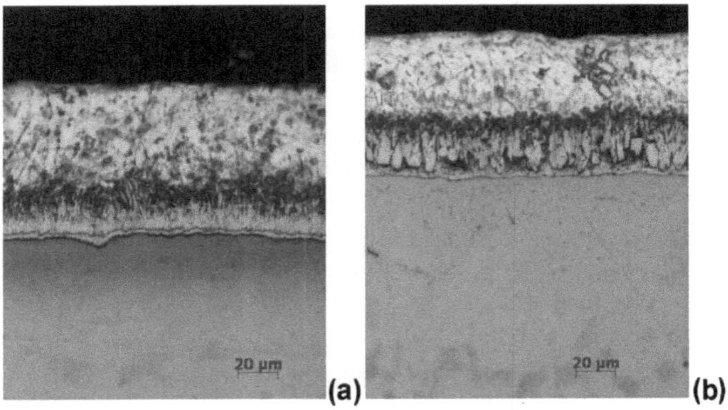

Figure 4.8 Coating after immersion time of 60 s: a) S235 steel (Si 0.012%), b) S355 steel (Si 0.196%).

diffusion of iron and further formation of the alloy coating is decelerated, on sample b) the relatively thin and inconsistent layer of phase δ or disintegrated crystals of phase ζ do not prevent diffusion of iron and the coating achieves a greater thickness with time (see also Section 5.4.7).

During low-temperature hot-dip galvanizing, zinc partly diffuses into the steel where it forms the primary solution α in the thin surface layer. On this substrate, on sufficient heating of the surface, the iron-zinc alloy phases δ and ζ immediately start to grow while their proportional representation in the coating changes with the immersion time.

The thickness and structure of the resulting coating depend on a number of other factors influencing the course of the metallurgical reaction. The section thickness of the material to be galvanized has a considerable impact on the thickness of the final

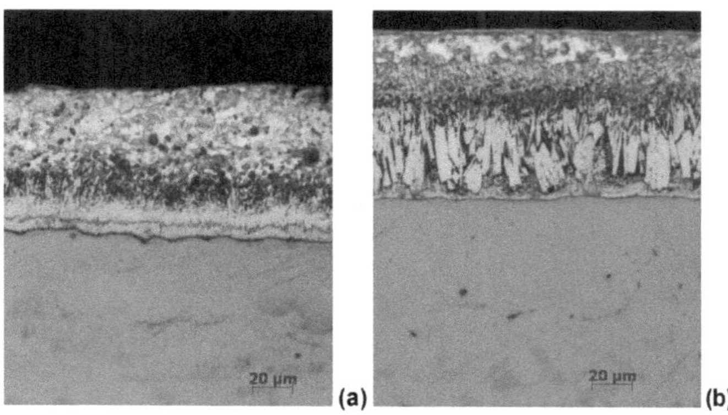

Figure 4.9 Coating after immersion time of 120 s: a) S235 steel (Si 0.012%), b) S355 steel (Si 0.196%).

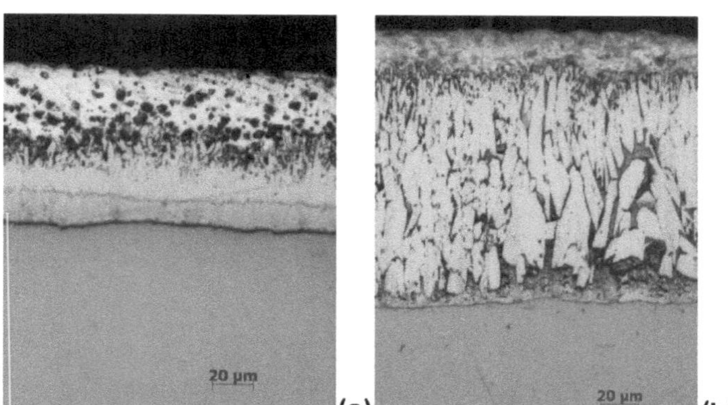

Figure 4.10 Coating after immersion time of 300 s: a) S235 steel (Si 0.012%), b) S355 steel (Si 0.196%).

coating. There is a general rule that on thin substrates the coating is thinner than on thicker substrates [36].

What is decisive for the structure of the applied coating is the chemical composition of the steel, especially the content of silicon [37] added to killed steel as a deoxidization agent. Also, the crystalline structure and quality of the steel surface, the method of its mechanical and heat treatment, and last, but not least, the presence of foreign substances in the surface layer of the substrate could influence the resulting properties of the coating. A combination of these factors leads to the conclusion that the coating produced using the batch hot-dip galvanizing process in commercial galvanizing plants exhibits a wide range of various morphological variations.

4.3 High-temperature galvanizing

High-temperature galvanizing is a process conducted at higher temperatures than 530°C (usually at 550°C), i.e., above the stability temperature of phase ζ, which therefore cannot be generated under these conditions (Figure 4.11). High-temperature galvanizing is generally associated with centrifuging (to remove all zinc adhering to the surface of galvanized parts after removal of the batch from the zinc bath) and rapid cooling in water. These measures eliminate equilibrium conditions for a peritectic phase transformation at which the mixture of δ phase crystals and the zinc melt could produce an undesired mixture of crystals of phase ζ and phase η during slow cooling. The high-temperature metallurgical process involves significant dissolution of iron that bonds with zinc, producing phase δ. The conditions are not favorable for

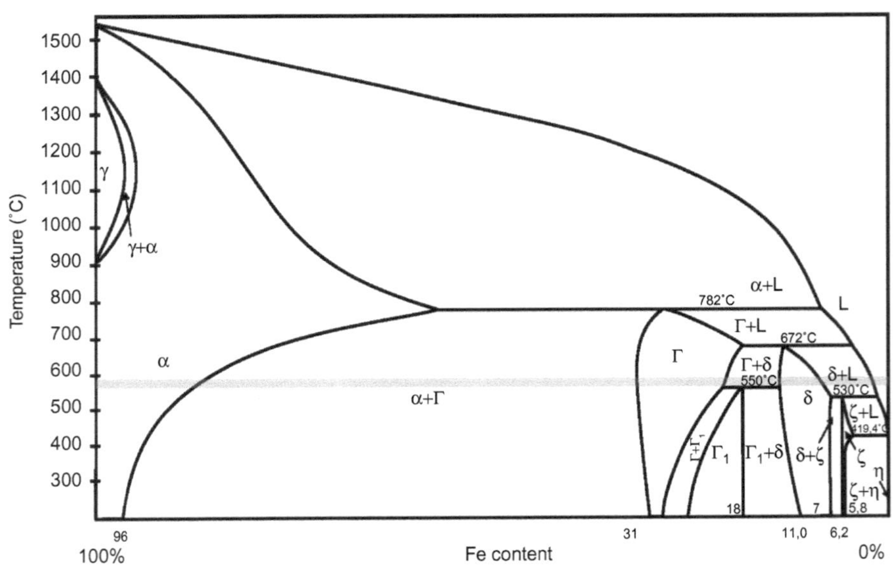

Figure 4.11 The gray strip in the diagram indicates the area of high-temperature galvanizing.

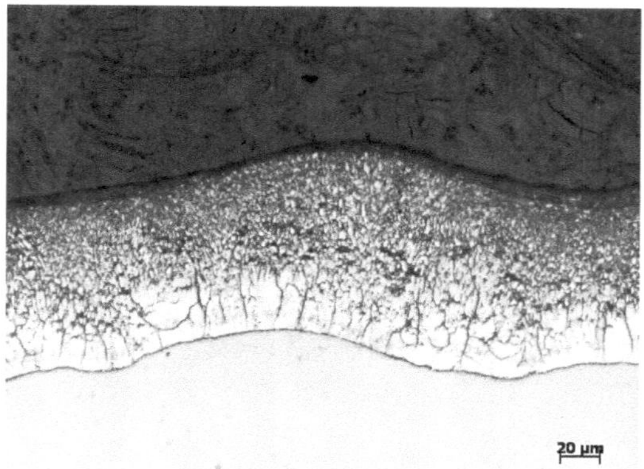

Figure 4.12 Coating on S355 steel, bath at the temperature of 550°C, immersion time 5 minutes, cooled in water.

generation of phase Γ, as within this technology the immersion time of the batch in the zinc bath is relatively short – generally not exceeding 100 seconds. During this time period, ferrite does not get sufficiently saturated with zinc so that the oversaturated solution α can be generated and the quick cooling in water mainly prevents segregation of phase Γ. Due to the quick cooling, phase Γ_1 cannot be segregated from phase δ either. In the alloy coating phase δ predominates, which is dense closer to the substrate, and further from the substrate there are fine crystals of phase δ overlayed with pure zinc (Figure 4.12). The thickness of coating applied by high-temperature centrifuge galvanizing is relatively even and the coating follows the surface of the galvanized part. Therefore, high-temperature galvanizing is preferably used, e.g., for coating of threaded fasteners (see also Sections 2.2.3 and 5.4.6).

Morphology of hot-dip galvanized coatings

5

The metallurgical reaction between iron and zinc is very sensitive to various external influences. Deviations in the chemical composition and the steel structure as well as galvanizing conditions lead to the occurrence of variable surface structures [38].

Note

All concentration data of admixtures or pollutants in this book are specified in mass units.

5.1 Influence of silicon (Si)

Oxygen is introduced into steel during its production process to reduce the content of undesired elements. The solubility of oxygen in steel lowers with decreasing temperatures. To prevent bonding of free oxygen with carbon, and so producing exogenous bubbles of carbon monoxide, which affect the characteristics of produced steel, steel must be deoxidized before ingots are cast. Elements with a higher affinity to oxygen than that of iron are used for this deoxidization. Examples of these are manganese, aluminum, or silicon. Silicon is a very efficient deoxidization agent. Its presence in steel during hot-dip galvanizing, however, has an unfavorable impact on the characteristics of the coating.

Dependence of the coating characteristics on silicon content in the galvanized steel was studied by Robert W. Sandelin and he came to the conclusion that on steel with a silicon content over 0.03% the thickness of the coating suddenly and abruptly increases. Later, John J. Sebisty continued Sandelin's research to discover that when the Sandelin range is exceeded (i.e., 0.03 to 0.12% by mass of Si content) the coating thickness temporarily decreases with the increasing content of silicon. With further increasing of Si concentration over approximately 0.25%, the coating thickness rises again (see also Section 5.1.2 and Figure 5.6).

5.1.1 Non-silicon-killed steel

From the point of view of coating, structural steels that have not been "killed" using silicon are generally recommended for hot-dip galvanizing. Coatings on such steel contains a substantial proportion of the alloy δ phase, which is characterized by close packed columnar hexagonal crystals. Their impermeable layer very efficiently

Hot-Dip Galvanizing of Steel Structures. DOI: http://dx.doi.org/10.1016/B978-0-08-100753-2.00005-7

separates the substrate from the zinc melt and relatively quickly decelerates the diffusion of iron and its reaction with zinc. On the layer consisting of δ phase there is a layer of relatively fine crystals of ζ phase arranged in the lawn form. Alloy layers of the coating are generally covered by a layer of pure zinc stuck to the surface and penetrating into the disintegrated crystals of ζ phase. Zinc coating with this structure exhibits good mechanical characteristics. The coating thickness on non-silicon-killed structural steel usually slightly exceeds the minimum values defined by the EN ISO 1461 standard. Extension of the immersion time in the zinc bath does not result in a considerable increase of the coating thickness on these types of steel.

As soon as the surface of parts of ferritic steel (which has very low silicon content) is sufficiently heated up, a relatively thin, continuous layer of δ phase is immediately created on it together with the primary layer of ζ phase, which is characterized by a relatively consistent parallel arrangement of close packed monoclinic crystals (Figure 5.1). Continuing diffusion of iron atoms first gives rise to the secondary layer of ζ phase separated from the primary layer by a perceivable boundary, while the thickness of δ phase does not change very much in this time interval (Figure 5.2). On some substrates, the presence of multiple layers of ζ phase created on each other can be rarely found (Figures 5.4 and 5.5). It is only after a

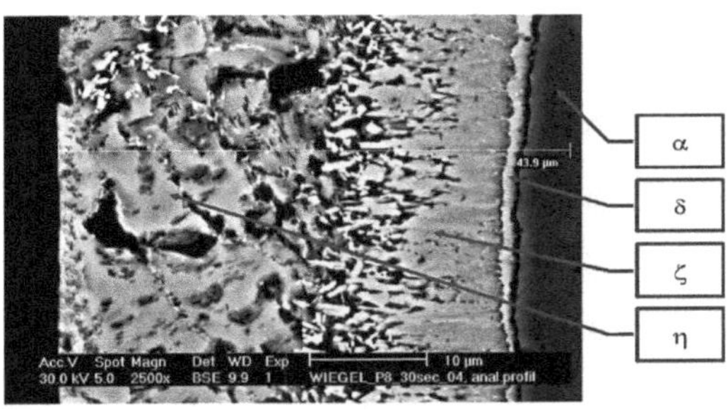

Figure 5.1 Coating on S235 steel with Si content of 0.003% after immersion time of 30 s in the zinc bath.

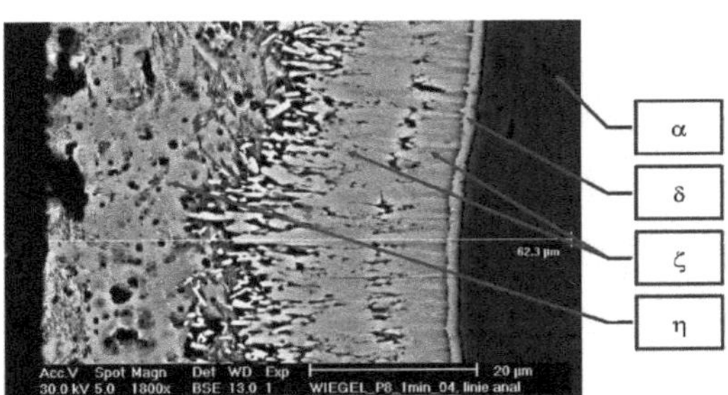

Figure 5.2 Coating on S235 steel with Si content of 0.003% after immersion time of 60 s in the zinc bath.

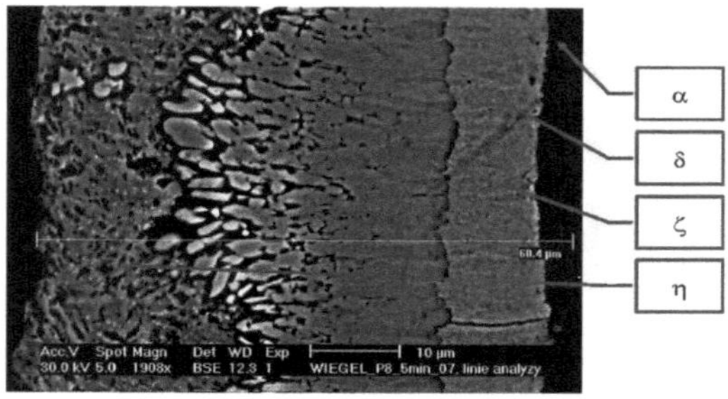

Figure 5.3 Coating on S235 steel with Si content of 0.003% after immersion time of 300 s in the zinc bath.

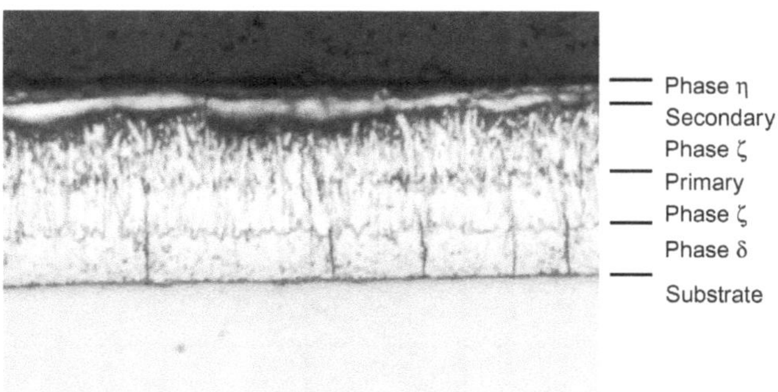

Figure 5.4 Two layers of monoclinic crystals of phase ζ laid on a layer of hexagonal crystals of phase δ.

sufficient dwell time in the zinc bath that the thickness of δ phase substantially rises and its matured layer represents a very efficient barrier, acting against the diffusion of both the metals. Therefore the metallurgical coating-forming process gets decelerated (Figure 5.3).

On some substrates both the layers of ζ phase combine to form a single layer and the original boundary between them disappears. In other cases, a distinctive double or multiple layer of ζ phase remains in the matured zinc coating (Figures 5.4 and 5.5). With the increasing distance from the substrate the crystals of ζ phase get disintegrated and are overlayed with zinc melt.

5.1.2 Silicon-killed steel

When silicon-killed steel is to be galvanized, the situation gets more complicated. Silicon influences the reactivity of iron with zinc in a fundamental way. It has an unfavorable impact on the coating characteristics. Coatings on these steel types can be thick and brittle, prone to mechanical damage as a layer of alloy phase ζ consisting of randomly arranged monoclinic crystals predominates in them. However,

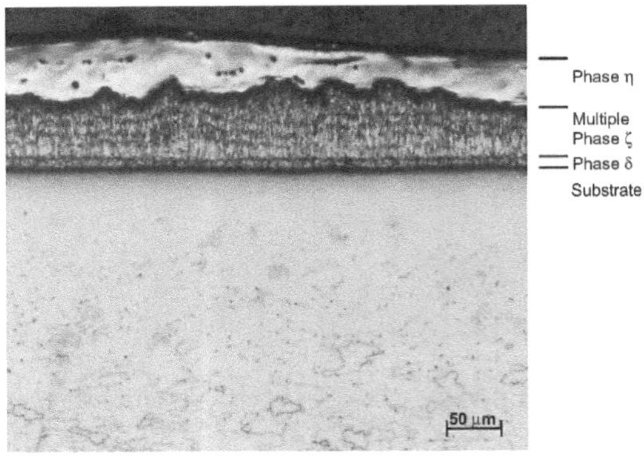

Figure 5.5 Multiple layer of phase ζ separated by a significant phase boundary from the layer of hexagonal crystals of phase δ.

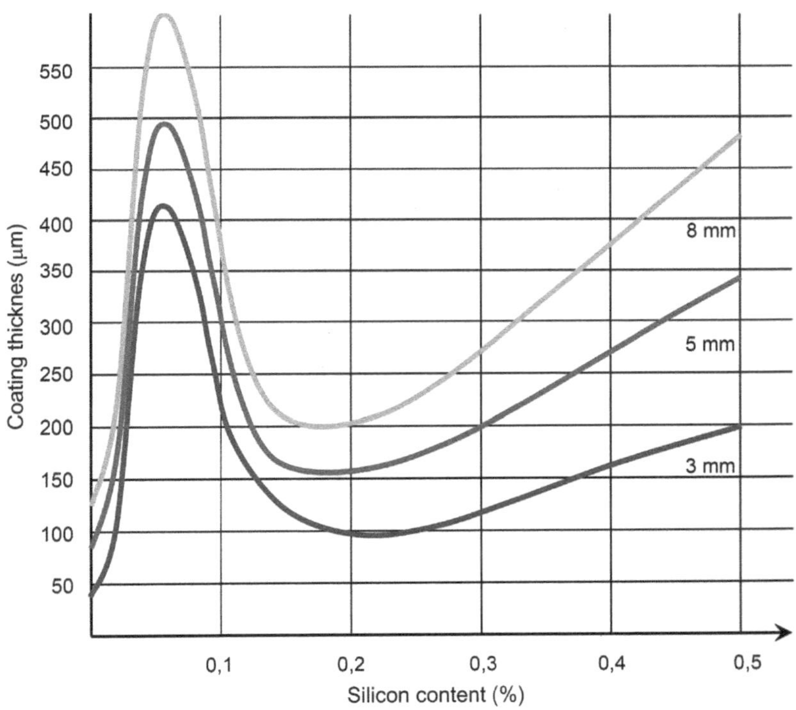

Figure 5.6 An example of the Sandelin diagram.

coating of silicon-killed steel may also have an acceptable thickness and structure if you manage to keep the content of silicon in steel within the limits defined by the "Sebisty range," i.e., between 0.15 and 0.25% Si.

The dependence of the thickness of the hot-dip galvanized coating can be graphically represented with the use of the Sandelin diagram (Figure 5.6). Figure 5.14

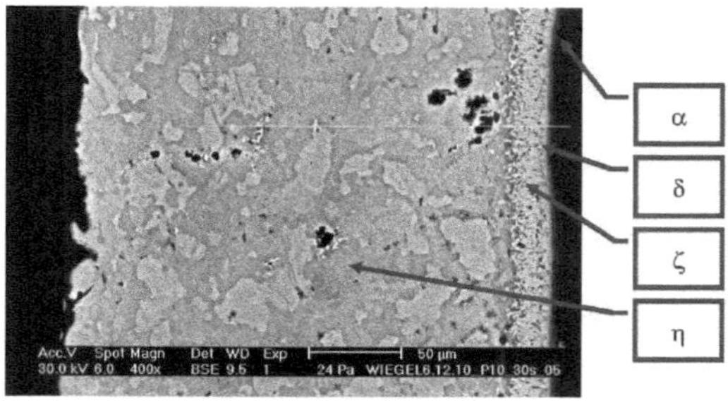

Figure 5.7 Coating on S355 steel with the Si content of 0.242% after immersion time of 30 s in the zinc bath.

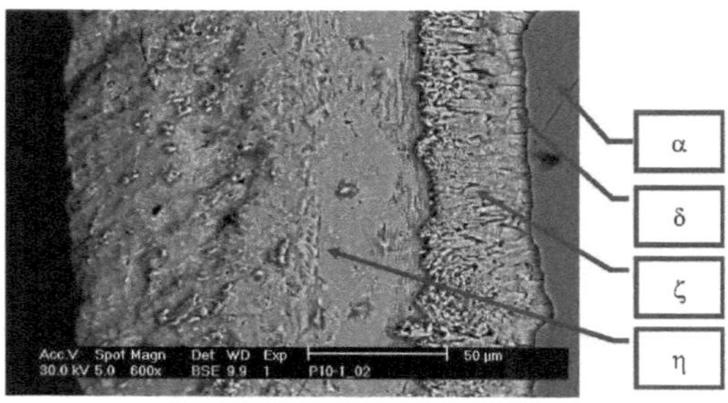

Figure 5.8 Coating on S355 steel with the Si content of 0.242% after immersion time of 60 s in the zinc bath.

shows a weldment of two different materials. The glossy tube of the column is made of a typical steel grade not killed by silicon while the sheet of the column foot is made of high-quality fine-grained silicon-killed steel. The figure clearly indicates that the coating of the foot gets a matte shade a few seconds after the removal already (all pure zinc was bonded to iron before solidification, producing an alloy coating layer) while the coating of the column tube keeps the glossy light shade of pure zinc.

On parts of structural silicon-killed steel a thin layer of phase δ is quickly formed on the part surface after its sufficient heating and this layer is covered by a layer of phase ζ, which at this moment consists of fine, randomly arranged monoclinic crystals (Figure 5.7). With the continuing immersion time of the part in the zinc bath an inconsistent layer of phase δ forms in the coating whose hexagonal crystals of variable length, arranged perpendicularly to the substrate, form columns penetrating into the ζ phase layer (Figure 5.8). Diffusion of iron continues and the ζ phase layer develops. After a sufficient immersion time in the zinc layer the originally fine crystals of phase ζ unite into compact units of variable size and with the continuing immersion time the thickness of their layer increases linearly (Figure 5.9). On such a substrate, diffusion of iron atoms is not prevented by a sufficient barrier of the δ phase layer and the metallurgical process of coating formation can continue (see also Section 5.4.7).

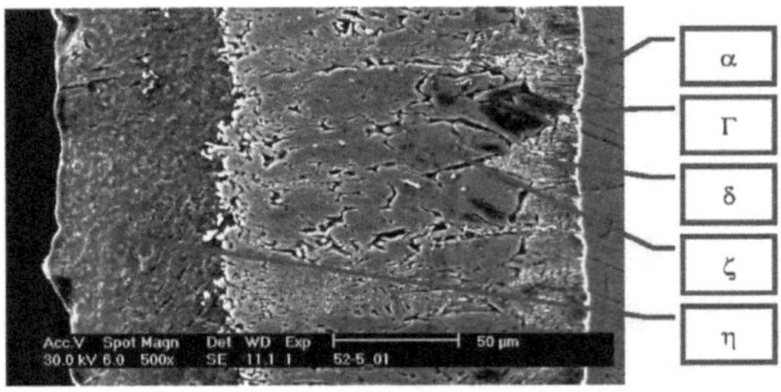

Figure 5.9 Coating on S355 steel with the Si content of 0.242% after immersion time of 300 s in the zinc bath.

Note

The Sandelin diagram only provides approximate information. Since the final thickness of the zinc coating depends on a range of other influences as, e.g., the thickness of the galvanized material, chemical composition of steel, zinc bath temperature, or immersion time, the diagram cannot be used to reliably determine the coating thickness. During batch galvanizing, the part is suspended at an angle with the maximum possible inclination of all the walls and edges so that excess zinc can better flow off its surface during the removal of the part from the bath. There is a principle for galvanized batches that the side of the part that is the first to get in contact with the melt should be the last to leave the zinc bath during withdrawal. On the top element of a galvanized part the coating thickness is lower than at the bottom end. Besides the immersion time and chemical composition of steel, the coating thickness is also materially influenced by the section thickness of the galvanized part.

Four basic structures of alloy coatings by silicon content in steel are shown in Figures 5.10–5.13.

Coating on steel with low silicon content, i.e., approximately up to 0.03%, is characterized by a structure with three distinctive layers of δ, ζ, and η phases (Figure 5.10). The alloy phases δ and ζ consist of fine, close packed crystals and are always covered by a pure zinc layer (phase η). The coating has a very good mechanical resistance and its thickness slightly exceeds the minimum values prescribed by the EN ISO 1461 standard.

Relatively thick coatings consisting mostly of solidified, mushy matter are formed on steel with silicon content within the Sandelin range (0.03 to 0.12%). The matter consists of fine crystals of phase ζ dispersed in pure zinc (Figure 5.11). The surface of such coatings is often affected by runs (see Section 6.1.1). It may exceptionally exhibit reduced adhesion caused by shrinkage porosities resulting from intensive diffusion of iron from the substrate.

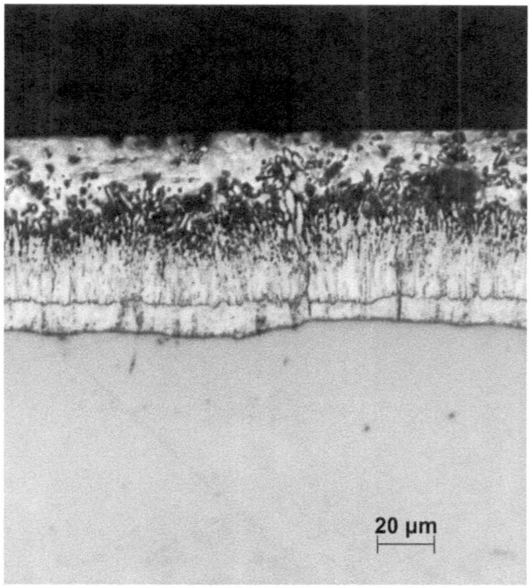

Figure 5.10 Coating on steel with Si content of 0.003%.

Figure 5.11 Coating on steel with Si content of 0.08%.

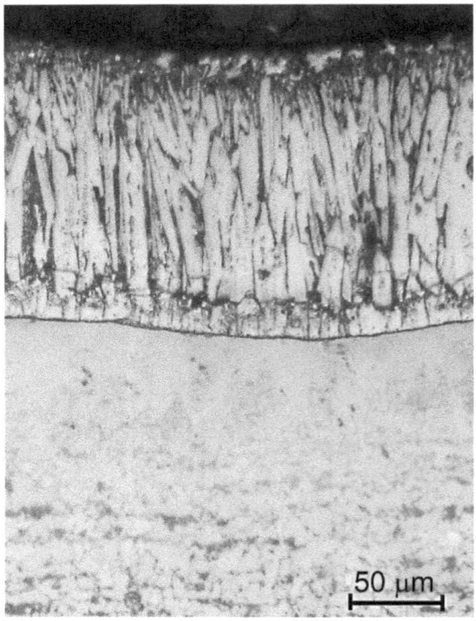

Figure 5.12 Coating on steel with Si content of 0.21%.

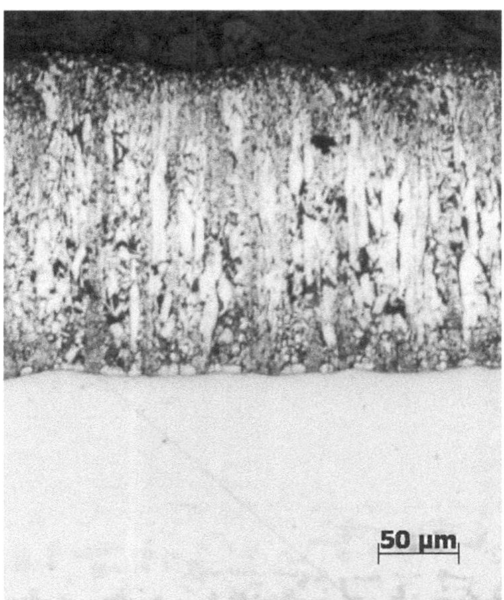

Figure 5.13 Coating on steel with Si content of 0.38%.

Figure 5.14 Combination of different steel grades.

A suitable compromise to achieve a zinc coating with acceptable mechanical properties on a silicon-killed substrate is to use a steel with a silicon content in the Sebisty range (0.15 to 0.25% Si). These coatings (Figure 5.12) are thicker and more brittle than coatings on non-silicon-killed steel. However, their mechanical resistance is acceptable and before application of organic coatings they can be successfully sweep-blasted.

Steel with high silicon content (above 0.28%) cannot be recommended for hot-dip galvanizing. Such steel coatings are very thick and brittle (Figure 5.13). They have very low mechanical resistance and even if they are subjected to very careful sweep blasting, they delaminate easily (see also Section 7.8.1).

5.2 Other influences of chemical composition of steel

The properties of a galvanized coating are influenced by some other elements [15, 39, 40] that the steel may contain as admixtures or as undesired impurities. The galvanizing result may also be affected by foreign substances present only in the surface layer of the substrate on which crystals of phase ζ preferentially grow during solidification of liquid metal after removal from the bath.

5.2.1 Influence of phosphorous (P)

A similar effect on the zinc coating formation to that of silicon with content within the Sandelin range has been proven for phosphorous. Its influence is observed in steel with a silicon content of up to 0.03% Si. To prevent generation of the Sandelin effect, the following condition should be met: $Si + 2.5\,P < 0.09$ [41].

5.2.2 Influence of sulfur (S)

An increased content of sulfur, which is generally added to free-cutting steel to improve its machinability, causes excessive dissolution of steel in the zinc melt [15]. Thus, during hot-dip galvanizing the geometry of parts made of steel that contains more than 0.15% of sulfur may change. In particular, threaded surfaces become non-functional in this case.

5.2.3 Influence of manganese (Mn)

Manganese at levels above 1.25% Mn support more closely packed arrangement of monoclinic crystals of phase ζ. Coatings on silicon-killed steel with this content of manganese generally exhibit a higher consistency and better mechanical properties.

5.2.4 Influence of carbon (C)

Structural steel with guaranteed weldability, which is galvanized most frequently, has a carbon content up to 0.2%. Carbon contents up to 0.3% have a relatively small impact on the reaction between iron and zinc [40]. Higher contents accelerate the reaction rate and result in a higher coating thickness. Free carbon precipitated at grain boundaries (e.g., in welds) may be the cause of hydrogen corrosion during pickling. Suitable alloying admixtures support the formation of carbides resistant to hydrogen in the trace quantity already.

5.2.5 Influence of aluminum (Al) in steel

Experiments conducted with samples affected by mixed structures (see Section 6.1.4) show that one of the possible causes of the occurrence of this phenomenon may be aluminum contained in steel. Similarly to silicon, aluminum is added to killed steel as a deoxidizing agent. At the same time, with regard to its high affinity to nitrogen, its presence increases the resistance of steel to aging. A possible aluminum surplus may remain in steel as free atoms or be bound in alloy phases of iron with aluminum (e.g., $FeAl_2$, Fe_2Al_5, or $FeAl_3$). During the hot-dip galvanizing process, free aluminum in the surface layer of the substrate takes priority in entering the reactions and gives rise to an inconsistent, very thin and at the same time dense iron-aluminum phase layer, which prevents iron diffusion where it occurs. The resulting iron-zinc alloy coating then exhibits considerable non-uniformity.

5.2.6 Influence of other elements

Chromium and nickel are contained in stainless steel, which is not usually galvanized. Additions of niobium, titanium, and vanadium are found in fine-grained steel. However, their contents in steel are insignificant and the influence on surface properties is negligible [40].

5.3 Influence of additions in the zinc bath

To improve the coating characteristics and to facilitate its application, some other metals are added to the zinc bath. The total contents of other elements in the zinc bath (except iron and tin) must not exceed 1.5% (by mass) in accordance with the EN ISO 1461 standard.

5.3.1 Aluminum (Al) in the zinc bath

Aluminum is added to the zinc bath to reduce its oxidation at the surface. In commercial galvanizing plants its content is generally maintained in the range of 0.003 to 0.005%. Higher aluminum contents cause formation of a thicker layer of aluminum oxide on the zinc bath surface, which cannot be dissolved by the flux applied onto the zinc parts. This leads to galvanizing defects. Aluminum is generally added to the zinc bath by adding a zinc alloy that has been purchased with a higher content of aluminum. Aluminum contained in the coating at even higher levels can increase its corrosion resistance. Baths with such a higher content of aluminum (usually 5 to 15%) required special conditions of coating application. They can only be met in galvanizing plants for continuous galvanizing.

5.3.2 Tin (Sn)

Tin added to the zinc bath in a concentration up to 1% supports formation of "zinc spangle." Galvanized coatings produced in baths with added tin are characterized by this distinctive pattern that enhances their decorative impression. Tin contained in the zinc bath in a concentration up to 1% supports reactivity of zinc with iron and increases the coating thickness in steel with a very low silicon content (up to 0.02%) (Figure 5.15). On the other hand, in silicon-killed steel the addition of tin in a concentration of approximately 0.5% Sn results in a reduction of the coating thickness (Figure 5.16). The impact of tin on reducing the coating thickness on silicon-killed steel may be supported by simultaneous addition of nickel to the zinc bath [40].

According to some sources [23], tin contents exceeding 0.1% may increase susceptibility to LMAC (Liquid Metal Assisted Cracking); therefore adding of tin to zinc baths is being abandoned at present [42] (see also Section 10.8).

Figure 5.15 Hot-dip galvanized coating on non-silicon-killed S235 steel in a zinc bath with Sn content of: a) 0%, b) 0.5%, c) 1%.

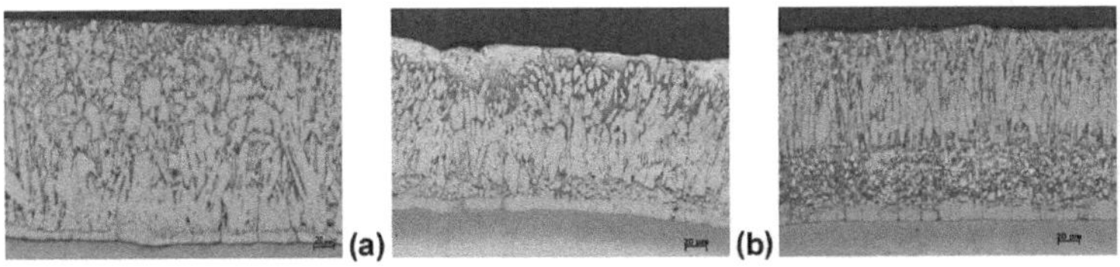

(a) (b) (c)

Figure 5.16 Hot-dip galvanized coating on silicon-killed S355 steel in a zinc bath with Sn content of: a) 0%, b) 0.5%, c) 1%.

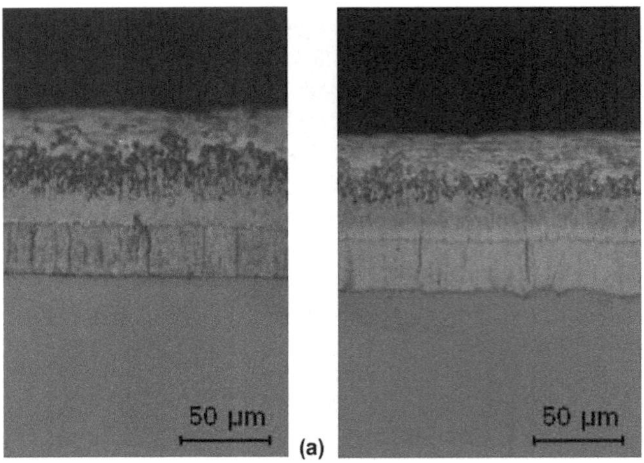

(a) (b)

Figure 5.17 Coating on steel with Si content of 0.01% galvanized in a bath: a) without Ni, b) with 0.055% Ni [35].

5.3.3 Nickel (Ni)

Addition of nickel to the zinc bath suppresses the Sandelin effect on steel with silicon contents between 0.03 and 0.12%. In a zinc bath with a nickel content of approximately 0.055% these steel grades behave in a similar way as steels that have not been killed by silicon. The coating from a nickel-containing bath is consistent, has an adequate thickness, and features very good mechanical characteristics. If the steel contains more than 0.12% silicon, nickel no longer has any influence on the coating properties [35].

The photos in Figures 5.17–5.19 show a comparison of coatings applied on steel with different silicon contents, namely in a zinc bath: a) without addition of nickel and b) with added nickel.

5.3.4 Lead (Pb)

Lead is added to the zinc bath to reduce the surface tension, reduce the zinc melt viscosity, and improve its flow off the surface of galvanized parts during their removal from the bath. In zinc bath with a temperature of 450°C lead dissolves up to the levels of approximately 1.2% and it is added to the bath in the pure form. It is convenient to

Figure 5.18 Coating on steel with Si content of 0.05% galvanized in a bath: a) without Ni, b) with 0.055% Ni [35].

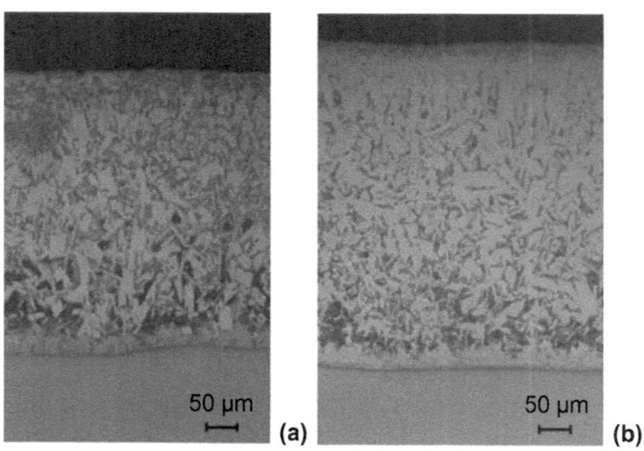

Figure 5.19 Coating on steel with Si content of 0.32% galvanized in a bath: a) without Ni, b) with 0.055% Ni [35].

operate zinc baths with a surplus of lead, which is concentrated on the bottom of the galvanizing tank due to its high specific weight. A several-centimeter thick layer of lead protects the tank bottom from mechanical damage and supports collection of dross for recycling. A problem is the regulatory constraints on lead due to its health effects.

5.3.5 Bismuth (Bi)

Given the toxicity of lead, some commercial galvanizing plants add a small quantity of bismuth to galvanizing baths as an alternative to lead to increase their fluidity [40]. Its content is maintained below 0.1%.

5.4 Influence of other factors

The properties of zinc coating applied by immersion in molten zinc are influenced by a number of factors or a combination of their effects. Setting optimum galvanizing parameters for all combinations of possible factors is difficult.

5.4.1 Substrate section thickness

The section thickness of the part to be galvanized has a fundamental influence on the thickness of the galvanized coating. EN ISO 1461 prescribes respective minimum coating thickness values for four specified ranges of substrate thickness. There is a rule that the coating thickness is smaller on thin-section substrates than on thick-section ones (see also Section 7.4.6).

5.4.2 Substrate structure

Carbon in galvanized structural steel does not have a significant effect on the final coating thickness. Coatings applied on ferritic steel or on ferritic-pearlitic steel are comparable. Accordingly, the texture of the galvanized steel does not have any impact on the galvanizing result. Disturbing the structure of the crystal lattice of ferrite in the surface layer of the substrate during hot-dip galvanizing releases the bonds between iron atoms and supports their diffusion and the boundary between the substrate and the zinc melt. This may lead to formation of mixed structures (see also Section 6.1.4), or the achievement of the required degree of saturation of the zinc melt with dross crystals is formed in the zinc melt surrounding the wall of the galvanized part under favorable conditions, which may stick to the coating in the powder form (see Section 6.1.3).

5.4.3 Mechanical processing of the substrate

Changes of the surface layer of the substrate caused by grinding or sand-blasting will change its profile and increase the contact surface of the substrate with the zinc melt. As a result, the coating on a material treated in this way achieves higher coating thickness than on a substrate with an as-rolled surface (Figures 5.20 and 5.21) [18].

Metal cutting of steel to be galvanized also influences the coating structure. The tool blade pressure during turning causes changes in the structure of the surface layer of steel, which is subsequently manifested in the coating structure by increased roughness (Figures 5.22–5.24) [18].

Figure 5.25 shows a different character of galvanized coatings as generated on a turned surface and inside a milled groove. The coating on the turned surface exhibits apparent scabs (see also Section 6.1.6) caused by damaging of the surface of the galvanized part due to incorrectly adjusted turning parameters.

Similar damage of the substrate as caused by metal cutting may be caused by excessive concentrated pressure in case of incorrectly adjusted parameters of a

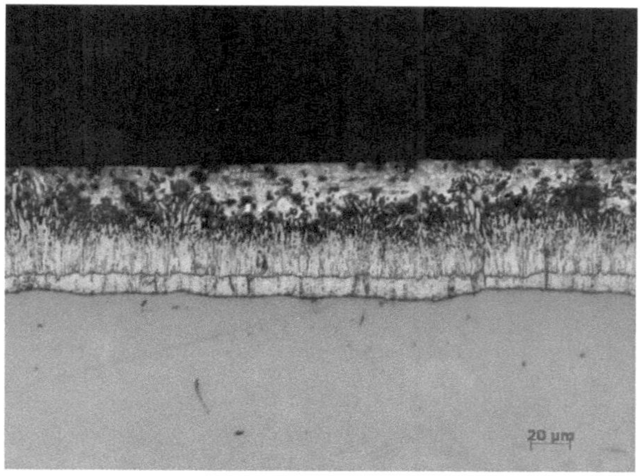

Figure 5.20 Coating on a sheet of steel S235 with rolled surface.

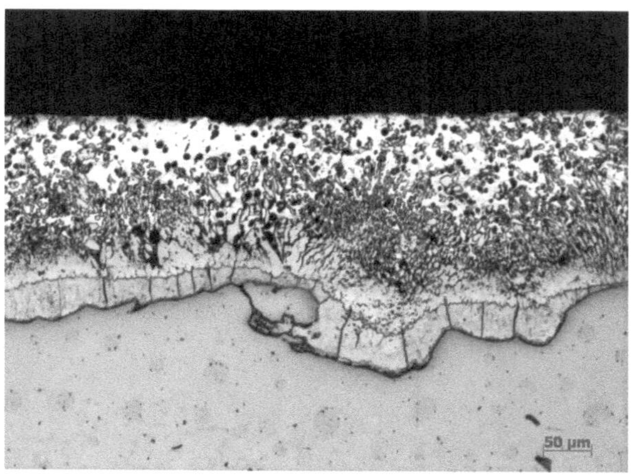

Figure 5.21 Coating on a sheet of steel S235 with sand-blasted surface.

Figure 5.22 Turned surface before galvanizing.

Figure 5.23 Turned surface after galvanizing.

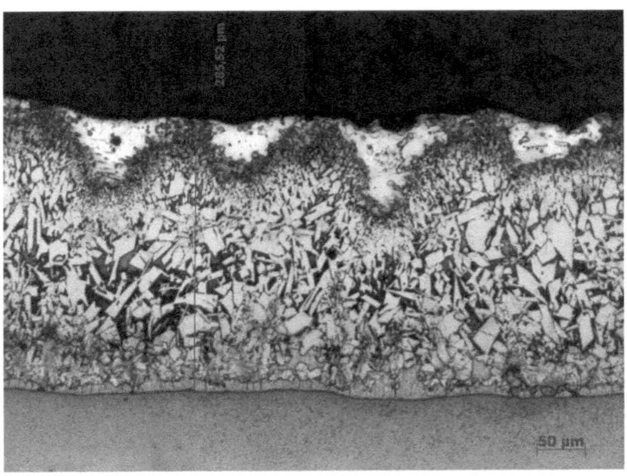

Figure 5.24 A metallographic sample of coating on turned surface shows perceivable traces of residual zinc corresponding to the lead of the helix of the turning tool.

cold forming operation. Figure 5.26 shows a galvanized part with a rough coating structure inside a rolled groove. The cause of the roughness is damage of the surface layer of the substrate during cold rolling (Figures 5.27 and 5.28) [18].

During rolling of steel, the surface layer may become contaminated by impurities, which may be impressed into the substrate by rolling (Figure 5.29). In such a situation, they are the cause of occurrence of mixed structures (Figure 5.30). The photo shows a considerably rough coating on a part of otherwise homogeneous

Figure 5.25 A different character of coatings produced on turned and milled surface.

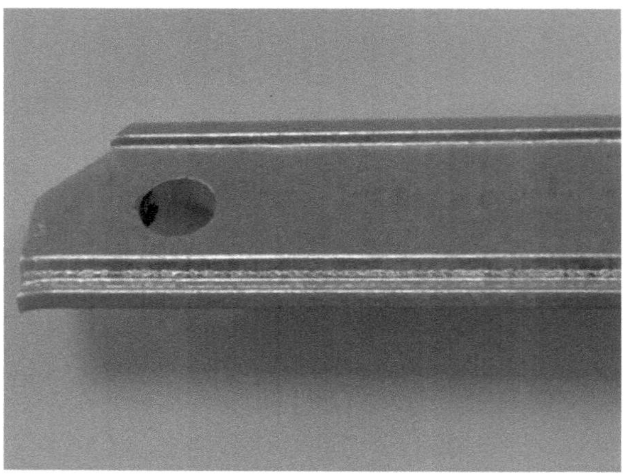

Figure 5.26 Rough coating inside a rolled groove after galvanizing.

material. In the primary production stage, impurities were rolled into the surface layer of the part of the surface of the steel sheet. After re-galvanizing this phenomenon is not likely to appear again (see also Section 6.1).

5.4.4 Thermal processing of the substrate

Annealing of steel produces scale at the expense of the surface layer of galvanized steel. Annealing at temperatures without recrystallization having already taken place (values below Ac_1) achieves recovery of the reinforced surface layer and grain refining. Removal of the contaminated surface layer and change of the steel structure lead to formation of uniform galvanized coatings without local anomalies (Figure 5.31).

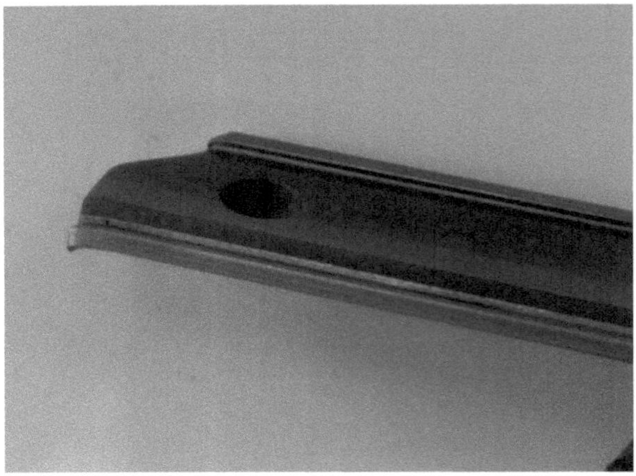

Figure 5.27 Glossy facet on the groove bottom before galvanizing.

Figure 5.28 Damage of the surface layer of steel in a rolled groove.

Figure 5.29 Rough coating on a contaminated sheet surface.

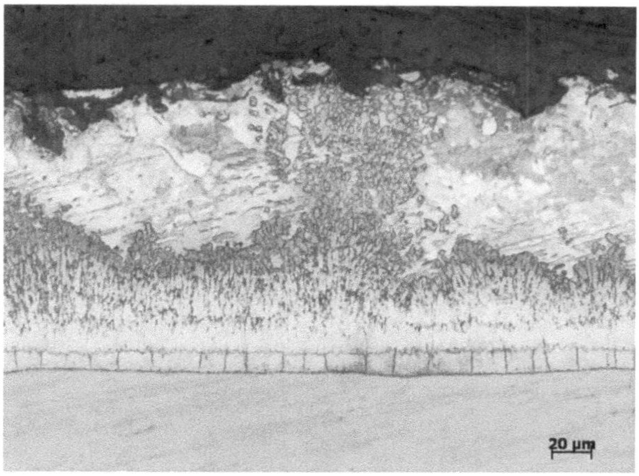

Figure 5.30 Rough coating structure found on a sheet with contaminated surface.

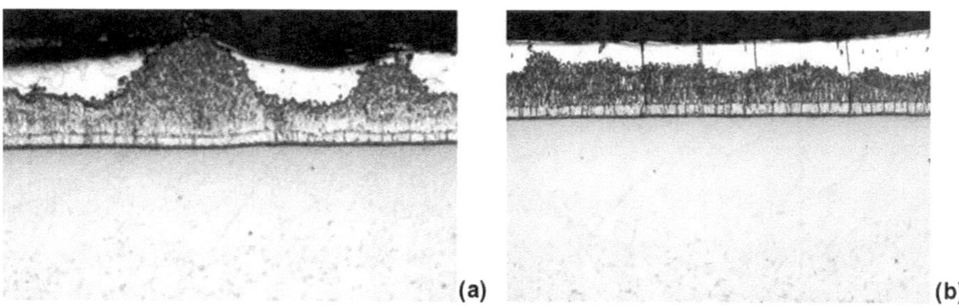

Figure 5.31 Influence of annealing: a) mixed structure on steel with a contaminated surface layer, b) coating structure on the same substrate that was shortly annealed at 600°C before galvanizing.

Due to the high energy demands of annealing, this process is not used to improve the zinc coating quality. The surface layer of steel can also be removed by using very long pickling times. However, harmful influences of the generated hydrogen must be taken into account (see Sections 3.6 and 10.1).

5.4.5 Substrate surface cleanliness

The zinc melt surrounding the surface of the galvanized part is partly saturated with iron that dissolves from the substrate. When a sufficient degree of saturation of the zinc liquidus with iron is achieved (at the zinc melt temperature of 450°C it is approximately 0.05%) crystals of phase ζ are formed in the zinc melt (see Section 6.1.2). They are preferentially formed on crystallization cores, which are foreign substances present in the melt. They can get there from the substrate surface [18]. Figure 5.32 shows the result of galvanizing of fencing made of closed hollow profiles that were operationally exposed for several years.

Figure 5.32 Impurities in cavities that cannot be removed by standard pre-treatment lead to galvanizing defects.

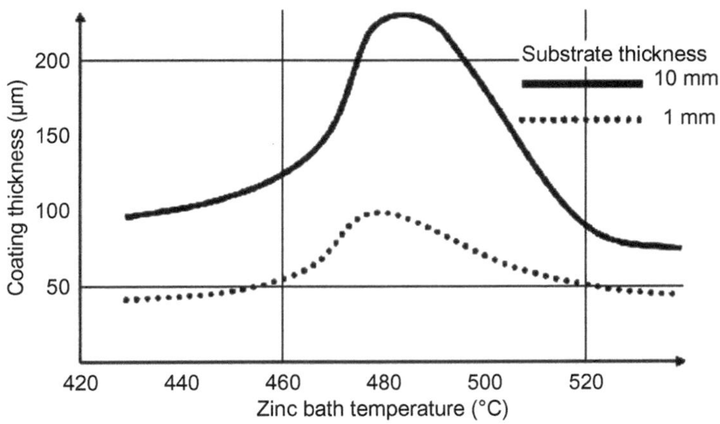

Figure 5.33 Temperature-dependence of the coating thickness, immersion time 5 min, Si content <0.03%.

5.4.6 Zinc bath temperature

The coating thickness for low silicon steel (with Si content up to 0.03%) is approximately directly proportional to the bath temperature in the interval between 440 and 480°C (for steel containing silicon in the Sandelin range, i.e., 0.03 to 0.12%, a similar proportionality is valid in a narrower interval – from about 440 to 460°C). Further increasing of the bath temperature causes a reduction of the coating thickness (Figure 5.33).

This dependence does not hold for steel containing silicon in the Sebisty range (0.15 to 0.25%), and a temperature increase in the interval of 440 to 470°C

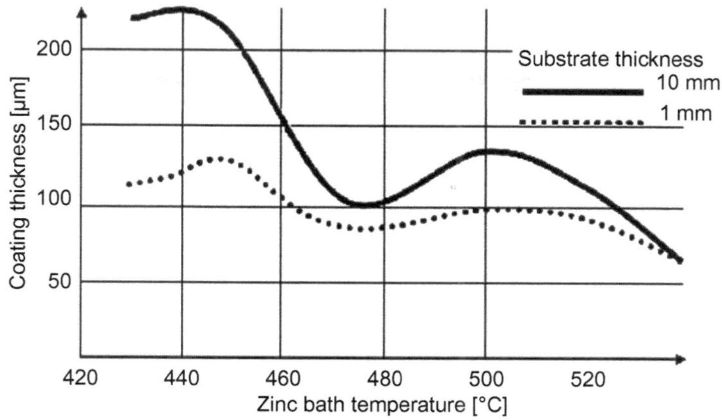

Figure 5.34 Temperature-dependence of the coating thickness, immersion time 5 min, Si content >0.12%.

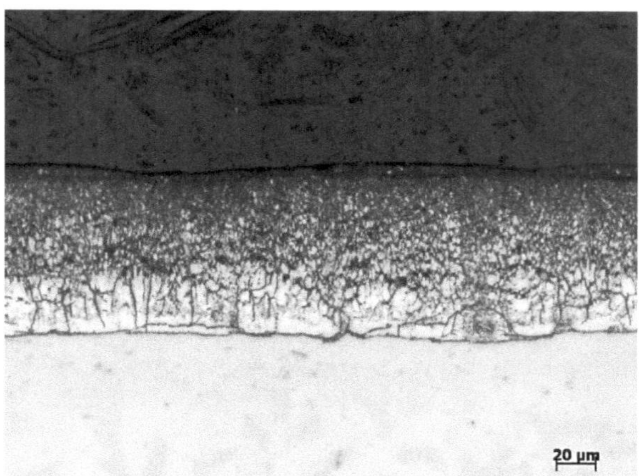

Figure 5.35 Coating on S355 steel applied with the immersion time of 5 min at 550°C.

conversely leads to a reduction of the coating thickness (Figure 5.34) [40]. At zinc bath temperatures above 530°C phase ζ is no longer formed (Figure 4.11) and phase δ becomes the predominant component of such coating regardless of the silicon content in steel, but its growth decreases with time (see also Sections 2.3 and 4.2).

The difference in the structure of coatings applied on silicon-killed S355 steel using the high-temperature and low-temperature centrifuge galvanizing process is shown in Figures 5.35 and 5.36.

Figure 5.37 compares the appearance of coatings applied onto sleeper bolts of S355 steel using the high-temperature galvanizing (matte – at the top) and low-temperature galvanizing process (glossy – at the bottom).

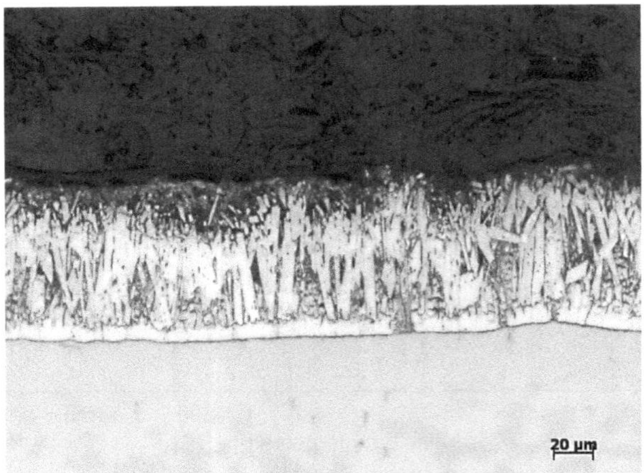

Figure 5.36 Coating on S355 steel applied with the immersion time of 5 min at 450°C.

Figure 5.37 Centrifuge galvanizing.

5.4.7 Immersion time

The immersion time only has a substantial impact on coating thickness with silicon content above 0.03%. In steel with very low silicon content (up to 0.03%) the dependence curve of coating thickness on the immersion time exhibits an approximately logarithmic relationship (Figure 5.38); further extension of the immersion time beyond the optimum value does not lead to any efficient increase of the coating thickness. The reason is priority formation of a consistent layer of hexagonal, close packed columnar crystals of phase δ, which causes a significant

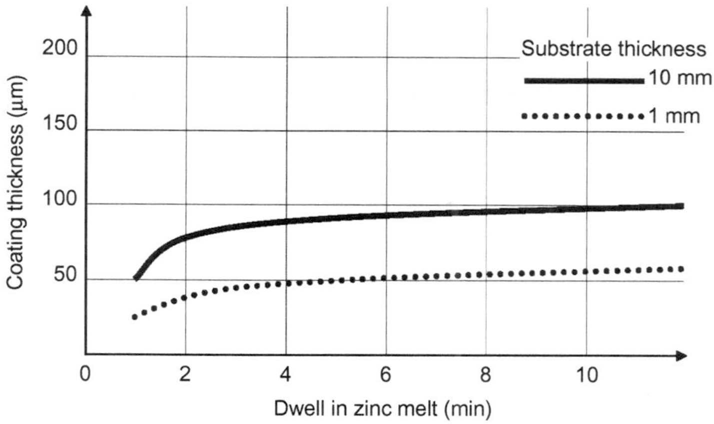

Figure 5.38 Dependence of the coating thickness on the immersion time, zinc bath temperature 450°C, Si <0.03%.

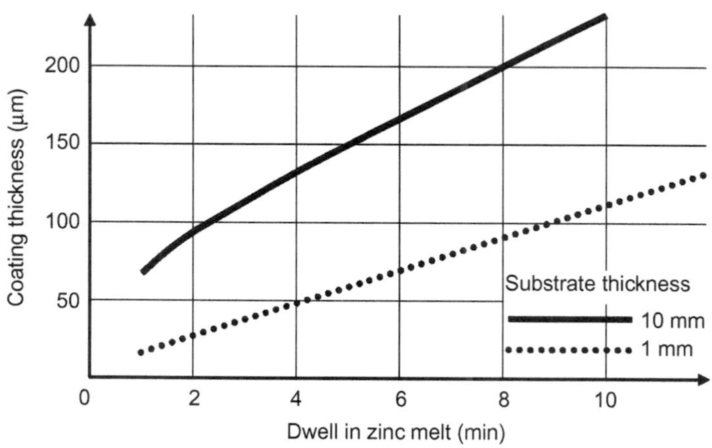

Figure 5.39 Dependence of the coating thickness on the immersion time, zinc bath temperature 450°C, Si >0.15%.

deceleration of iron diffusion, and the coating thickness growth gets slower. On steel with silicon content above 0.15% phase δ is formed in a thin, sometimes inconsistent layer and the diffusion of both the elements continues at the same rate. The coating thickness grows with an approximately linear dependence on the immersion time (Figure 5.39).

Figure 5.40 Coating on steel with Ø 50 mm with Si content of 0.012% after 48 hours' dwell in the zinc bath.

Figure 5.41 Coating on steel with Ø 50 mm with Si content of 0.316% after 48 hours' dwell in the zinc bath.

Note

Long immersion times always result in the formation of brittle coatings. Figure 5.40 indicates that coating applied with a long immersion time in the zinc bath spontaneously delaminates even on steel with a very low silicon content. After the same immersion time the sample in Figure 5.41 made of silicon-killed steel exhibited a fatal substrate loss with the formation of a large mass of iron-zinc alloy.

Figure 5.42 Color shade difference of the coating on the continuous surface of a steel part.

5.4.8 Cooling conditions

Diffusion of iron and zinc continues even after the removal of the batch from the zinc bath until the temperature of the parts drops below approximately 200°C when it is considerably decelerated. Especially in the case of reactive steel with silicon content in the Sebisty range and a higher section thickness, all zinc covering the coating surface is consumed to form iron-zinc alloy phases during cooling and the coating surface takes on a dull matte appearance. On very massive parts this phenomenon can already be observed in the course of a few seconds after removal from the zinc bath. Figure 5.42 shows a distinctive difference in the coating color on a 100 × 100 × 8 mm profile made of silicon-killed steel. A part of the coating in the place of faster cooling is covered by residual pure zinc giving the part a silverish glossy shade while at the other end of the bar, where the cooling was slower, all the zinc is transformed to iron-zinc alloy phases (see also Section 5.1.2 and Section 6.2.2).

Phenomena on galvanized coatings

6

6.1 Coating roughness

The thickness and structure of the coating produced by batch galvanizing depend on a large number of factors that influence the progress of the metallurgical reaction. The chemical composition of the steel to be galvanized, especially the amount of silicon added to "killed" steel as a deoxidizing agent, has a considerable influence on the reaction. The structure and nature of the steel surface, its section thickness and last, but not least, the presence of foreign substances on the surface or in the surface layer of that substrate can influence the coating properties. A combination of these factors means that coating produced in commercial galvanizing plants exhibits a wide range of appearance and morphology [18]. In practice, two basic structures are most frequently encountered. On non-silicon-killed steel (Figure 6.1) a fine consistent structure is formed whilst a rougher, more random, structure is often formed on silicon-killed steel (Figure 6.2) (see also Section 4.2).

6.1.1 Runs

In "Sandelin steels" silicon in a concentration between 0.03 and 0.12% causes an increased rate of release of iron atoms from the substrate and their diffusion into the zinc melt surrounding the article. When a certain degree of saturation of the zinc melt with iron is achieved, fine crystals are formed and a very viscous, mushy matter (Figure 6.4) is generated, which sticks to the surface of the galvanized article [18] and is generally also covered by pure zinc. During removal of the article from the zinc bath this mushy matter then causes runs to occur (Figure 6.3). Some galvanizing plants manage to eliminate these runs by adding nickel to the zinc bath, which suppresses excessive dissolution of iron in zinc. A milder form of bulking of the coating on the Sandelin steels is "orange peel" effect.

6.1.2 Lumps

Lumps (Figure 6.5) in the coating are formed from dross crystals sticking to the surface [18]. Unlike the zinc coating, which is produced by a diffusion process, these dross crystals are formed on foreign crystallization cores by transformation under equilibrium conditions. A precondition for such a transformation is a sufficiently high concentration of iron dissolved in the zinc bath. If the temperature of the zinc melt surrounding the galvanized article decreases, mainly due to heat

Hot-Dip Galvanizing of Steel Structures. DOI: http://dx.doi.org/10.1016/B978-0-08-100753-2.00006-9

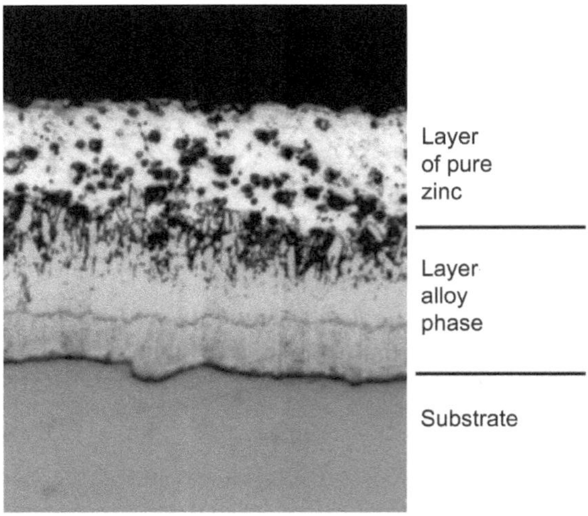

Figure 6.1 Coating on non-silicon-killed steel.

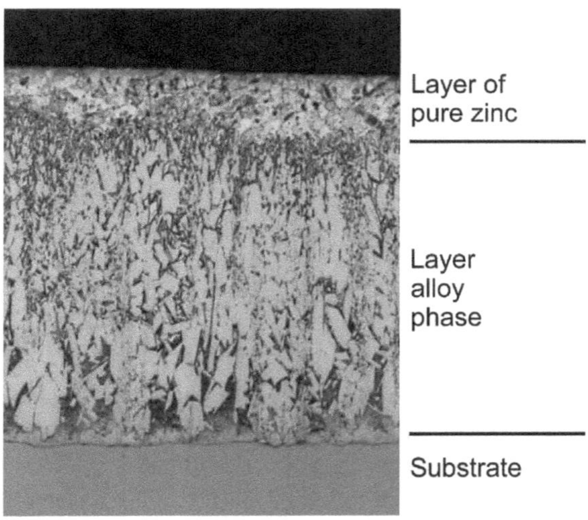

Figure 6.2 Coating on silicon-killed steel.

transmission to the galvanized material, some foreign particles (retained pickling solution, contaminated steel surface, etc.) may serve as crystallization cores where dross crystals can grow relatively quickly (Figure 6.6).

If some structural elements of a galvanized article are in a horizontal position during removal from the bath, free dross crystals may get caught on such a surface and then remain encapsulated in the layer of adhering pure zinc. Prevention mainly involves cleanness of cavities in the parts to be galvanized (see also Section 5.4.5).

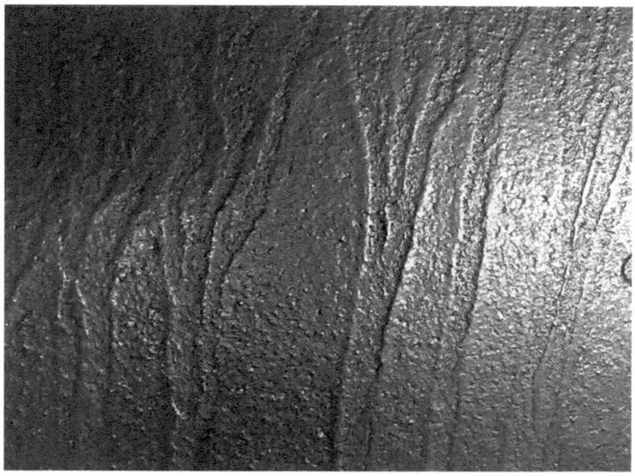

Figure 6.3 Coating affected by runs.

Figure 6.4 Mushy matter forming coating on steel with an Si content in the Sandelin range (consisting of fine crystals of phase ζ dispersed in pure zinc).

Figure 6.5 Deposited hard zinc lumps.

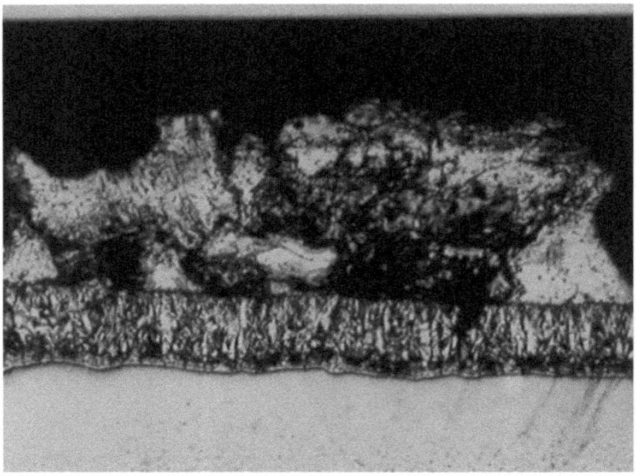

Figure 6.6 Section through a deposited lump.

Figure 6.7 Coating affected by grits.

6.1.3 Grits

Grits (Figure 6.7) are a phenomenon that is similar to the occurrence of lumps in the coating [18]. It usually occurs on sheet products of non-silicon-killed steel and is manifested as small protrusions resembling fine sand sealed in the coating (significantly finer than lumps). On a metallographic sample, monoclinic crystals of hard zinc (phase ζ) sealed in the pure zinc layer (Figure 6.8) can be found in the coating structure. The occurrence of these small crystals depends on sufficient saturation of the zinc melt with iron released from the substrate. Crystals of phase ζ can be formed under equilibrium conditions from a liquidus that is saturated with iron, on crystallization cores represented by atoms of other elements that are released from the substrate. These crystals stick to the surface of the part during its removal

Figure 6.8 Dross crystals sealed in the coating.

Figure 6.9 Example of a mixed structure.

from the zinc bath. In commercial galvanizing plants these equilibrium conditions are difficult to identify in advance and to reliably change.

6.1.4 Mixed coating structures

Non-homogeneous substrate

Galvanizing an article with a smooth and consistent surface that, however, exhibits local variations in structure and chemical composition of the steel may consequently give the impression that material suffering from a strong corrosion attack has been galvanized (Figure 6.9). Local differences of reactivity of the substrate with zinc cause the applied coating to have a so-called mixed structure [40] (see also Section 5.4.4).

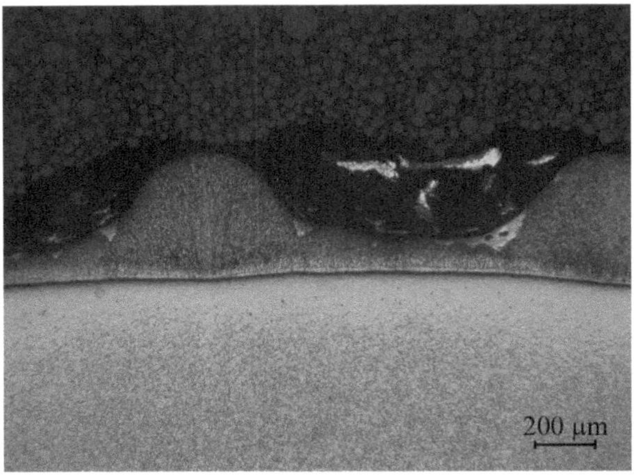

Figure 6.10 Metallography of a mixed structure.

Figure 6.11 Corrosive disruption of substrate.

A section through the coating very clearly shows places with a coating structure corresponding to silicon content in the Sandelin range that alternates with places exhibiting a coating structure formed on steel with a low silicon content (Figure 6.10) [18]. Adding nickel to the zinc bath may suppress occurrence of such mixed structures. After repeated galvanizing (after first removing the coating in a stripping bath) the mixed structure defect is not likely to be observed again.

Corrosive disruption of substrate surface

A specific case of occurrence of a rough structure is galvanizing of materials attacked by corrosion where after previous hot rolling the last operation performed was cold rolling. Corrosive attack (Figure 6.11) of the surface layer of such a

Figure 6.12 Raised pattern of thin galvanized sheet with the surface disrupted by corrosion.

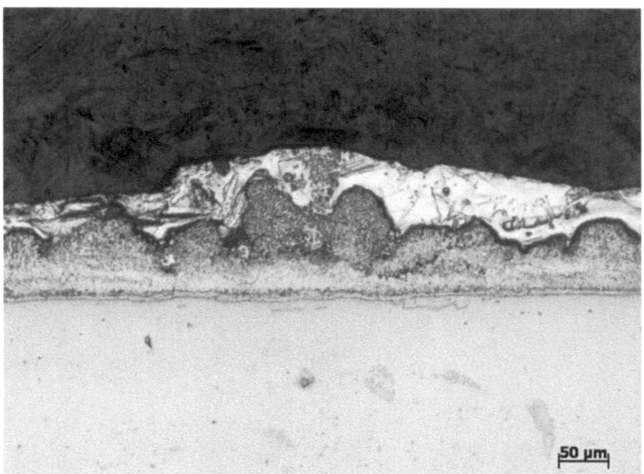

Figure 6.13 Mixed structures on a thin sheet with a cold-drawn surface attacked by corrosion.

substrate causes local defects of the crystalline lattice in the reinforced surface ferrite layer and disruption of the mutual bonds of iron atoms. During the metallurgical reaction of hot-dip galvanizing this supports easier release of iron atoms from the substrate, which promotes reactivity of steel with zinc. Mixed structures are formed in the coating (Figure 6.13) characterized by a raised pattern in the coating (Figure 6.12).

6.1.5 Laps

Forged or hot-formed semi-finished products may be affected by surface defects that are difficult to discover visualy – laps in the steel surface. However, during

Figure 6.14 Coating on a hot-formed tube.

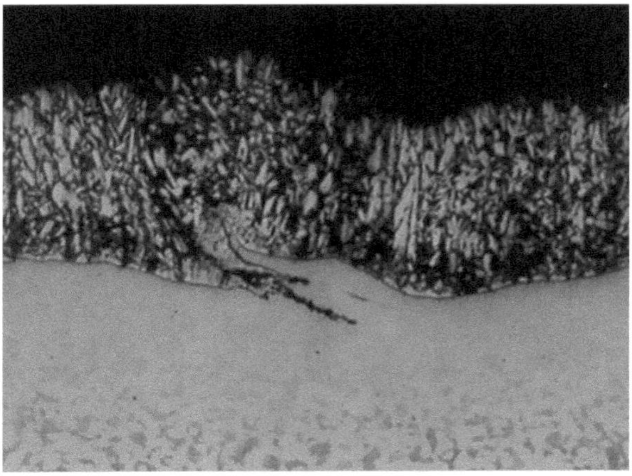

Figure 6.15 Lap on substrate steel.

hot-dip galvanizing they are considerably exaggerated (Figure 6.14). The laps get detached in the shape of needles or flakes (Figure 6.15) [18]. This is a very undesirable coating defect as there is a risk of physical injury.

Subsequent treatment by abrasion is virtually impossible, as the detached needles have steel cores and regrinding of the surface will expose them. Due to the reliable action of cathodic protection, there is no risk of corrosion of such exposed steel under normal exposure conditions in the atmospheric environment, but the service life of an anti-corrosion system affected this way is reduced. The EN ISO 1461 standard does not allow absence of zinc coating within acceptance tests. This problem may be solved by application of an organic coating. Parts designed to be hot-dip galvanized should not be made from these semi-products.

Figure 6.16 Scabbing of a pipe.

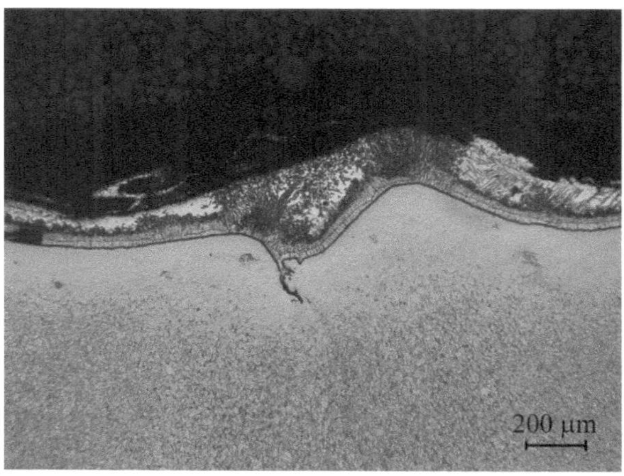

Figure 6.17 Surface disruption of the substrate.

6.1.6 Scabbing

Scabbing (Figure 6.16) is characterized by uneven places arranged in regular rows [18]. It originates from a disrupted surface of the substrate metal. A cause may be a failure to follow the operation process of the primary production, e.g., insufficient lubricant dosing during cold-drawing or calibration. This phenomenon may even be highlighted by the presence of silicon in the steel. Surface micro-cracks (Figure 6.17) cause zinc to react with iron on a free surface differently than in the constrained space of the cracks, and the zinc coating on the substrate grows in an irregular way. This is a defect whose occurrence cannot be predicted in normal practice (see also Section 5.4.3).

6.1.7 Shear steel

Unlike mild steel produced in the 19th century using a melting process at temperatures above 1600°C, the production of shear steel [11] consists of fining of dough-like crude iron in a reduction atmosphere in finery forges using the bloomery or puddling process [11]. Shear steel is characterized by a large number of material layers mutually interconnected with blacksmith welds. The material is strongly anisotropic to fibrous (Figure 6.18).

Between the individual fibers of shear steel there are gaps that may be filled by corrosive products of iron after long-term exposure in a corrosive environment. In the case of these materials it is often difficult to achieve the pure metallic surface required for application of zinc coating with the use of common chemical pretreatment processes in the galvanizing plant. The pickling solution penetrates into the gaps in the substrate, which causes coating defects, and during the metallurgical hot-dip galvanizing process the steel may even disintegrate (Figure 6.19).

Figure 6.18 Anisotropy (fibrous character) of shear steel.

Figure 6.19 Hot-dip galvanized historical grating of shear steel.

6.1.8 Foam

Foam (Figure 6.20) may be formed in the coating of steel containing silicon in the Sandelin range due to the escape of hydrogen that got into the steel during pickling. If at the moment the mushy matter of the mixture of the fine crystals of the phase dispersed in the zinc melt starts to solidify after removal of the galvanized part from the zinc bath, a large quantity of hydrogen escapes. This hydrogen may recombine and form bubbles (Figure 6.21) that cause foaming of the solidifying mushy substance (see also Section 6.1.1).

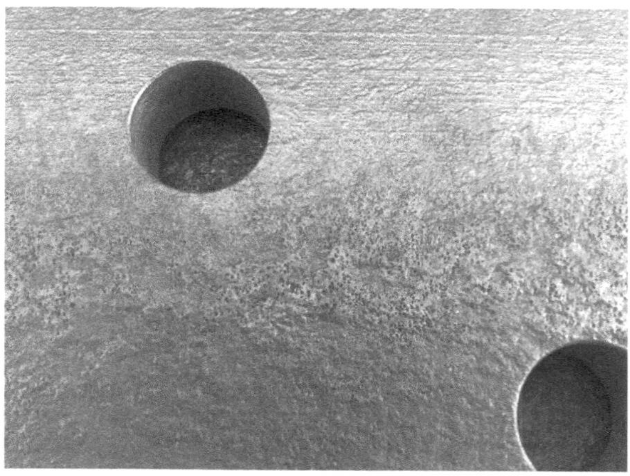

Figure 6.20 Foam in the galvanized coating formed on steel with Si content in the Sandelin range.

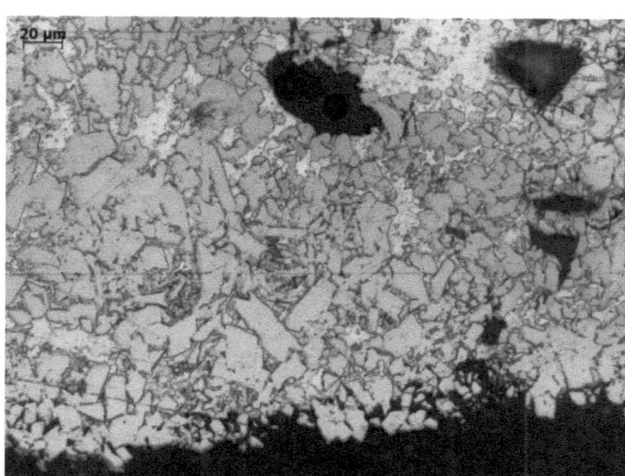

Figure 6.21 Hydrogen bubbles in coating affected by foam.

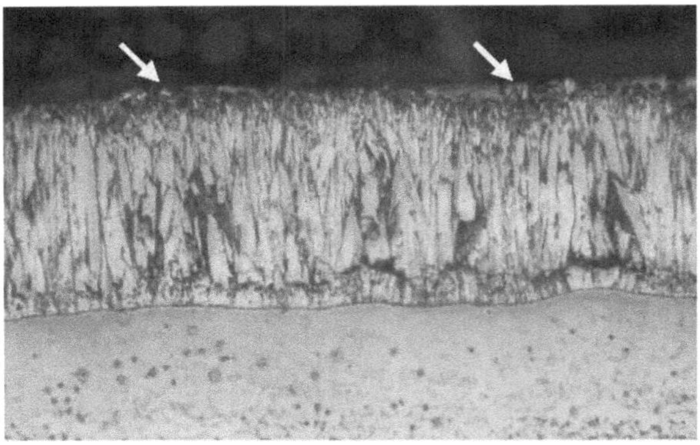

Figure 6.22 Continuous residual zinc.

6.2 Color shade

The aim of hot-dip galvanizing is to apply an alloy coating on the substrate that is produced during the immersion of the batch in a zinc bath through a metallurgical reaction between iron and zinc. When a galvanized part is removed from the zinc bath, a thin pure zinc layer always adheres to its surface, which persists in some coatings and not in others. Alloy coating covered by pure zinc has a light glossy color while alloy coating consisting exclusively of alloy phases has a matte and dull coating. The cause of this phenomenon is the fact that diffusion processes may continue in the coating until its temperature falls below approximately 200°C. After removal of the part from the zinc bath, any zinc stuck to the surface may partly or completely react with iron diffusing from the substrate, generating intermetallic phases. A coating on reactive substrates exclusively consists of alloy phases; pure zinc is not present. Sometimes the surface of a galvanized part may simultaneously have areas covered by residual pure zinc (Figure 6.22) and alloy-only areas.

The presence of a thin layer of pure zinc on the zinc coating surface does not have any influence on its corrosion resistance; coatings consisting of alloy phases only and coatings covered by a pure zinc layer are equal from the point of view of anti-corrosion protection performance.

6.2.1 Zinc spangle

A pattern on galvanized coating that has the shape of irregularly distributed triangles or quadrangles is called zinc spangle (Figure 6.23). It is caused by the flaky crystalline composition of zinc. Sharply delimited geometric patterns are the boundaries of these crystals. During hot-dip galvanizing in commercial galvanizing plants, zinc spangles are only created where there are equilibrium conditions for the occurrence of this phenomenon.

Figure 6.23 Zinc spangle.

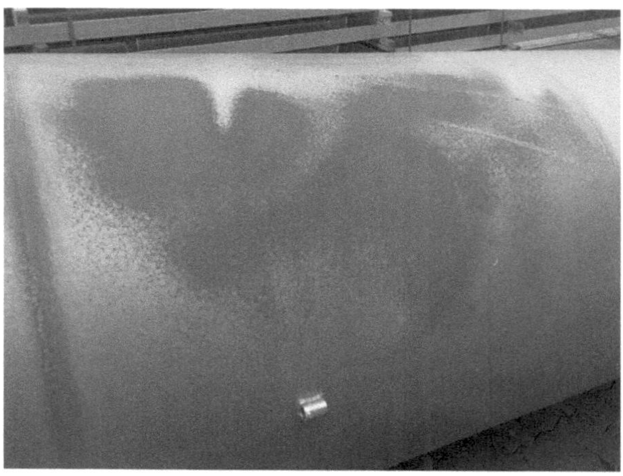

Figure 6.24 The charred spots phenomenon caused by a locally increased content of Si.

6.2.2 Heat spots

Matte enclosed areas with a gradual transition to glossy coating are called heat spots [43]. This term is derived from one of the possible causes of this phenomenon. Diffusion processes in the coating continue after removal of the galvanized part from the zinc bath until the temperature of the part drops below about 200°C. If the article is suspended on a jig near another article, the accumulated heat may cause local deceleration of cooling. The same effect may also be caused by an internal structural element, e.g., a massive partition inside a tube, as shown in Figure 6.24. While the surrounding coating is covered by a layer of pure zinc, in a slowly cooled place this zinc is consumed during the generation of matte iron-zinc alloys due to continuing diffusion. Another condition for the occurrence of two

Figure 6.25 Residual pure zinc on a coating part.

Figure 6.26 Non-homogeneous silicon content – dispersed.

different coating colors on the article surface may result from its dimensions. The tubular pole in Figure 6.25 was exposed to radiating heat of the zinc bath at the bottom side (on the right) during its removal from the bath while the top side (on the left) was cooled more quickly. Zinc is present on the side exposed to quicker cooling (see also Section 5.4.8).

6.2.3 Non-homogeneous substrate steel composition

The same effect as in the case of heat spots is caused by a locally increased concentration of silicon in the steel [43]. In a location of concentrated silicon content, the reaction between iron and zinc runs more quickly and the adhering pure zinc is consumed to create alloy phases immediately after emerging from the zinc bath. On the stringer in Figure 6.26, the time of temperature drop below 200°C after removal from the zinc bath corresponds exactly to the time when zinc sticking to surface

Figure 6.27 Non-homogeneous silicon content – locally concentrated.

Figure 6.28 Color difference on posts after galvanizing.

was being transformed into alloy phases. On one part of the surface it managed to form alloy coating by diffusion while the remaining part of the stringer is covered by pure zinc. In the case of the watering can in Figure 6.27, radiation of heat from an adjacent item can be excluded as a cause, as it was galvanized manually using the wet process. In this case, the cause of the matte stain on its surface is a local variation of the composition or the substrate structure.

6.2.4 Combination of different substrates

If steels of different quality grades are combined in a welded part, each substrate can react differently with zinc and result in a different color shade of the coating on different elements of the weldment (Figures 6.28–6.30).

Figure 6.29 Color difference on posts after several weeks' exposure.

Figure 6.30 Weldment of steels of different quality grades.

6.2.5 Marbling

Marbling is caused by local absence of residual pure zinc on an alloy coating (Figure 6.31). The marbling pattern is manifested on the coating of some substrates, usually steels with the silicon content in the Sebisty range, and may have the character of light netting (Figure 6.32) or light islands on a dark, matte base (Figure 6.33).

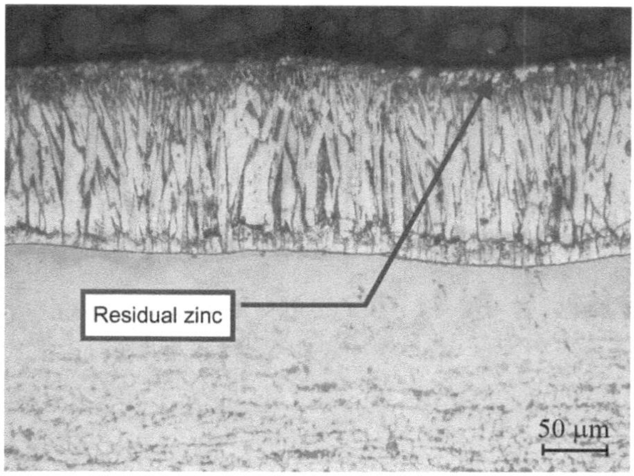

Figure 6.31 Marbling is caused by islands of residual zinc on alloy coating.

Figure 6.32 Marbling on freshly applied coating.

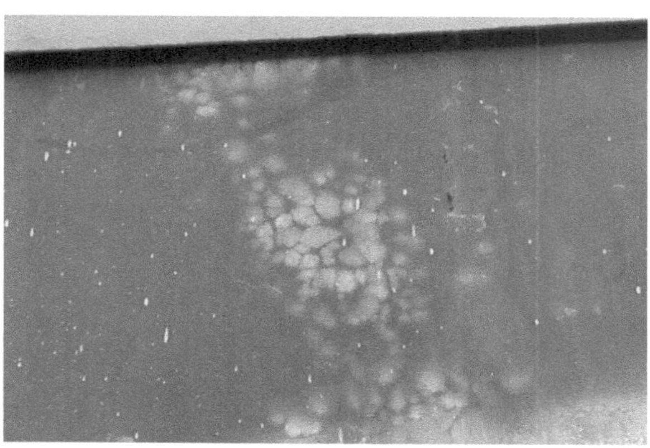

Figure 6.33 Marbling after a longer exposure time.

6.2.6 Patina

The light and glossy galvanized coating does not keep its color shade throughout its useful life, but if it is exposed in a corrosive environment in the course of a few months it will become covered by protective patina, which mostly consists of zinc carbonate produced by the transformation of zinc oxide, zinc hydroxide, or zinc hydrogen carbonate under equilibrium conditions, the transformation being supported by the atmospheric carbon dioxide (see Chapter 8).

Zinc carbonate is a gray material that is insoluble in water and it gives its dull color shade to hot-dip galvanized parts (Figure 6.34). The coating of articles that exhibited a zinc spangle after galvanizing will change in appearance after a longer exposure due to reaction of the surface layer with the atmosphere as the straight outlines of individual crystals become rounded (Figure 6.35).

Figure 6.34 Patina.

Figure 6.35 Zinc spangle after a longer exposure.

Figure 6.36 Darkening of the galvanized coating due to refacing by using silica-based abrasive.

6.2.7 Grinding of welds

To achieve high homogeneity and quality of the weld metal, the welding wire is characterized by relatively high silicon content acting in the liquid weld metal as deoxidizing reagent. Grinding of welds leads to the contamination of the substrate surface around the weld with silicon saturated metal. The grinding also causes a disruption of the crystal lattice in the surface layer of steel. Both of these factors are the cause of increased reactivity of iron with zinc. The zinc coating on such areas may take a matte gray color. It is free of residual pure zinc on the surface (Figure 6.36). Refacing of weld joint by using silica-based abrasive flap discs contributes to the coatings' darkening effect.

6.2.8 Trapped flux

Ashes produced by burning of flux during immersion of the batch in the zinc melt are usually washed out onto the surface. Sometimes, especially in case of insufficient drying, ashes may stick in a corner or to the wall of a galvanized article.

Unburnt flux contains aggressive components that attack the zinc coating. It should, therefore, be removed from the surface of galvanized articles in the galvanizing plant. After subsequent wetting, these residues cause a local attack of the coating (Figure 6.37). With regard to their considerable thermal inertia, massive parts can only be removed from jigs and cleaned when they have cooled down sufficiently. During this intermediate period, corrosion products of zinc cause a local change of its color shade on the coating. Places under trapped flux residues are significantly darker after their removal and give the impression of absence of zinc coating. Presence of a sufficient metallic layer can be usually verified using a coating thickness measurement device (Figure 6.38).

Figure 6.37 Trapped flux in a corner.

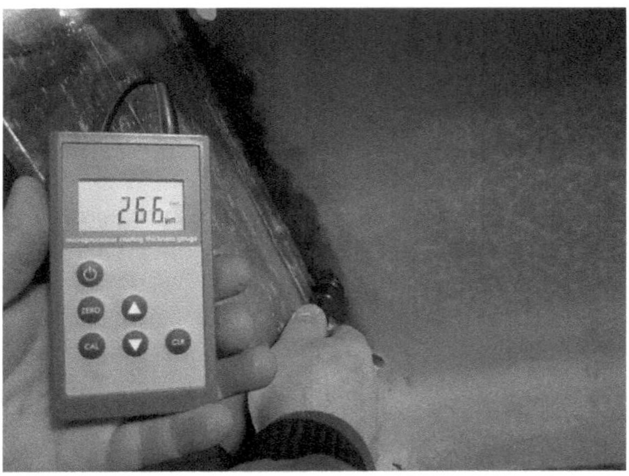

Figure 6.38 Coating thickness in the place of an imprint of trapped flux.

6.3 Coating on thermally cut surfaces

Oxygen, plasma, or laser cutting techniques are conveniently used to cut semi-products of structural steel due to the high productivity and economy of these methods. Articles designed to be hot-dip galvanized often contain thermally cut pieces whose cut surfaces are not treated in any other way (Figure 6.39).

Article 6.4 of the standard EN ISO 14713-2 declares that oxygen, laser, or plasma cutting changes the composition and structure of steel not only in the cut area but also in the thermally affected area. It is difficult to achieve the minimum prescribed coating thickness on cut surfaces, and the coating adhesion is often reduced. According to article 6.2.3 of EN ISO 1461, coating thickness must not be

Figure 6.39 Thermally cut surface covered by steel burning products.

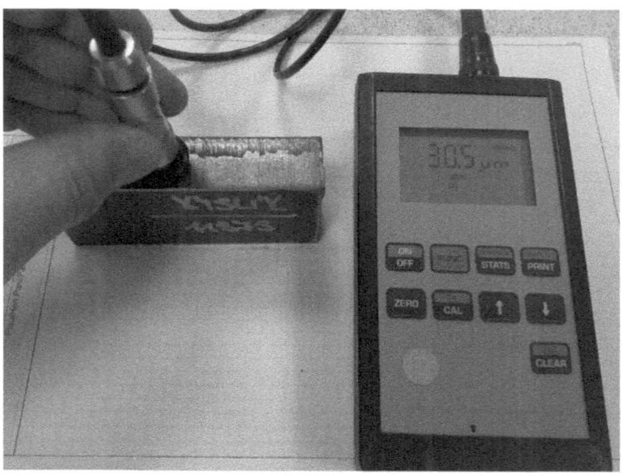

Figure 6.40 The surface after oxygen cutting of steel is covered by a non-magnetic layer of a foreign substance consisting of metal burning products and residual austenite.

tested on the surfaces of thermal cuts and at a distance of less than 10 mm from their edge. Figure 6.40 indicates that measurement of the coating thickness on the thermally cut surface using a magnetic method does not provide an objective result as before the application of the coating the relevant thickness of the non-magnetic layer can be determined here.

The oxygen cutting of steel is based on the principle of metal burning in an oxygen stream. An important precondition for oxygen cutting is that the ignition temperature of the base material and, at the same time, the melting temperature of the produced oxides must be lower than the melting temperature of the base material. During cutting, a sufficient amount of heat to heat steel to its ignition temperature is generated, which is approximately 1100°C, i.e., above the temperature of phase

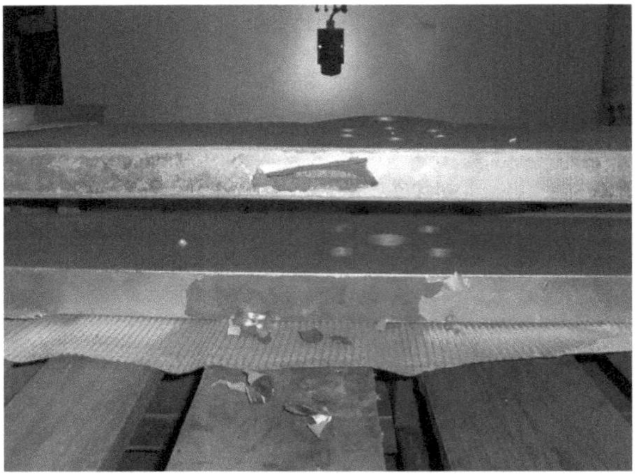

Figure 6.41 Delamination of a pure zinc layer on the scaled surface of a burnt cut.

transformation of ferrite to austenite, but still below the melting temperature of steel. The range of the thermally affected area of the material that is hit by the phase transformation of iron α to iron γ depends on the cutting conditions. Behind the cut, in the thermally affected area, the metal is quickly cooled by the surrounding unheated material, while at the same time martensitic transformation is going on inside it, which will stop when the cooling rate decreases. Therefore, a certain share of residual paramagnetic austenite remains in the surface layer. At the same time, due to combustion, the base material is being depleted of some elements in the surface layer of the substrate, e.g., of silicon [44].

Although oxygen cutting of steel should be exclusively done by burning the base metal without melting, its heating up above the melting temperature cannot be reliably excluded. The surface of a thermally cut edge of structural steel is always covered by a layer of metal that has passed, in a uncontrolled way, through all phase changes from the ferritic α structure through the austenitic γ and ferritic δ structure to liquidus and back. The cut surface is covered by a layer of scale – iron oxides (wüstite and hematite). If hematite prevails in them in a continuous layer or there is another unspecified ceramic coating, it is difficult to achieve a pure metallic substrate by pickling. The flux-activated, scaled surface of steel is wettable in the zinc bath and during hot-dip galvanizing, pure zinc sticks to it (Figure 6.41), which is characterized by low adhesion to the substrate (see also Section 3.2).

6.3.1 Thermally cut edges

The surface of the galvanized coating on thermally cut areas is characterized by the presence of pure zinc, which provides it with a light glossy color shade. The applied coating follows the surface roughness achieved by thermal cutting (Figures 6.42 and 6.43).

Figure 6.42 Coating on a thermally cut sample of S235.

Figure 6.43 Coating on a thermally cut sample of S355.

On the samples (Figures 6.44 and 6.46) the zinc coating applied on the thermally cut surface exhibits, regardless of the silicon content in the substrate, a continuous layer of relatively fine and close packed crystals of phase δ on which there is an approximately equally thick layer of close packed needle-like crystals of phase ζ covered by a layer of stuck pure zinc.

The alloy part of the coating is characterized by a structure corresponding to coating of ferritic steel with a trace content of silicon (sometimes its thickness does not achieve the minimum value prescribed by EN ISO 1461). A comparison to the coating structure created on a rolled surface of the same substrate is provided by Figures 6.45 and 6.47. Due to frequent occurrence of tangential cracks

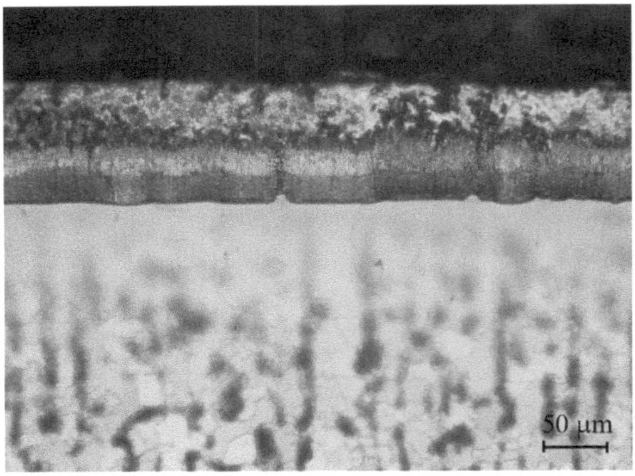

Figure 6.44 Coating – thermally cut of S235 steel.

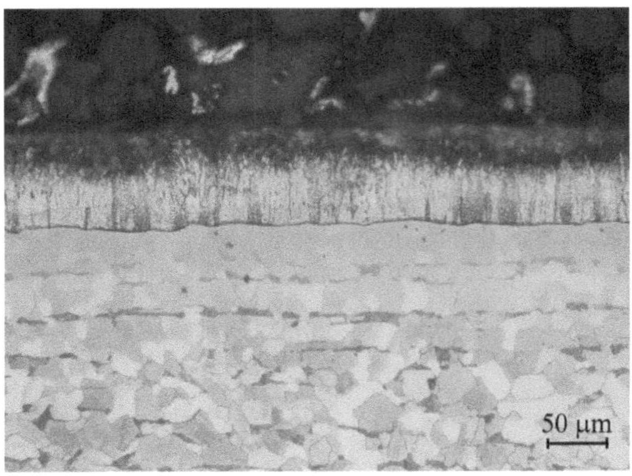

Figure 6.45 Coating – natural surface of S235 steel.

Figure 6.46 Coating – thermally cut of S355 steel.

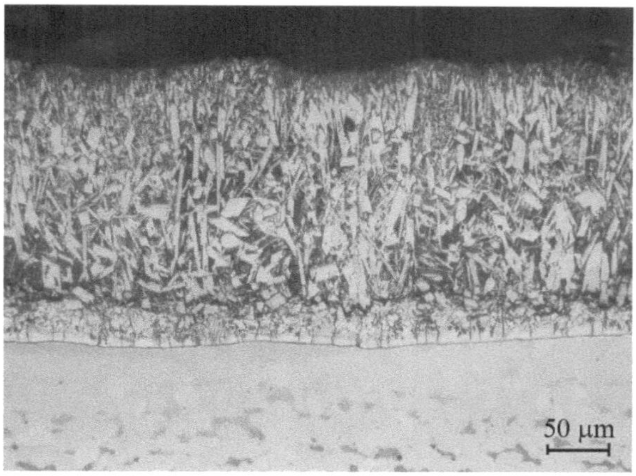

Figure 6.47 Coating – natural surface of S355 steel.

Figure 6.48 Tangential crack in the coating.

(Figure 6.48) the coating often exhibits reduced adhesion. Tangential cracks already occur due to different thermal expansion coefficients of the coating alloy and the basis steel in the stage of coating growth during the immersion of the galvanized part in the zinc bath. A crack formed this way may sometimes pass into the stage of primary delamination of coating (Figure 6.49) when a layer of the originating coating gets detached from the substrate to such an extent that it enables penetration of zinc melt into the resulting gap (see also Section 7.8.3). There, a secondary coating layer is formed, which, however, warps due to mutual shifts of the non-adjacent coating and substrate.

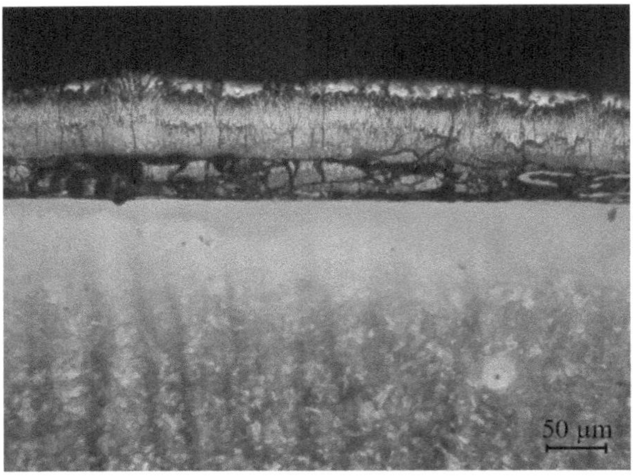

Figure 6.49 Primary coating delamination.

Figure 6.50 Coating delamination on the edge of a thermally cut of steel C10E.

6.3.2 Edge adjacent to the thermally cut surface

Due to thermal cutting, the edges of the cut are the most changed parts in terms of substrate structure and composition as well as surface contamination. The edge is affected by a step change of material characteristics of the basis steel. These are important factors influencing the coating adhesion. The performed metallographic analysis indicates that a galvanized coating is most affected by the presence of cracks reducing coherence and adhesion of the coating on the edges of thermal cuts (Figure 6.50).

Tests of hot-dip galvanized coatings and assessment of their quality

7

7.1 Order acceptance by the galvanizer

In terms of a commercial supply chain, surface treatment by hot-dip galvanizing carried out in a commercial galvanizing plant is execution of work on an article or component provided by the purchaser to the galvanizer. The galvanizer is obliged to execute the work with expert care; the galvanizer is especially obliged to notify the purchaser of any unsuitable characteristics of the article if the galvanizer is able to determine this unsuitability by applying appropriate expertise. The purchaser is obliged to deliver the article to the galvanizer in a condition that enables high-quality fulfillment of the work. If unsuitable aspects prevent proper implementation of the work, the galvanizer is obliged to interrupt the implementation unless an instruction is received from the purchaser that they insist on execution of the work having been informed of any unsuitable aspects. A prerequisite for high-quality application of the surface treatment is that the purchaser meets the principles of proper design and production of articles intended to be hot-dip galvanized (see the principles defined in Chapter 9).

Before confirmation of the order, the surface treatment contractor performs an input inspection, the objective of which is to identify possible defects in the articles that would prevent the execution of the work to the quality standards as defined by EN ISO 1461. Guidelines for selection of the anti-corrosion system and instructions for proper design and production of articles intended to be hot-dip galvanized are contained in the technical standards EN ISO 14713-1 and EN ISO 14713-2 (see Chapter 11).

The most frequent defects in articles to be hot-dip galvanized are incorrectly positioned or insufficiently sized or missing drain or vent holes. The surface and cavities of articles must not contain contaminants – residues of welding slag that could not be dissolved by typical chemical pre-treatment operations (see Chapter 3). Incorrectly designed overlapping joints and porous welds of poor quality are the cause of staining of the coating by corrosion products of iron seeping from the gap if the galvanized part is exposed in a wet environment. The usual quality standard cannot be guaranteed on untreated surface of thermally cut surfaces (see Section 6.3). Rigid structures face the risk of deformation or cracking by the thermal

Hot-Dip Galvanizing of Steel Structures. DOI: http://dx.doi.org/10.1016/B978-0-08-100753-2.00007-0

load of the hot-dip galvanizing process. Parts whose dimensions exceed the working dimensions of the galvanizing bath cannot be galvanized ("double-dipping" should only be applied in exceptional situations). The galvanizer, as an expert, should inform the purchaser about any risks that can be predicted and that can lead to poor quality execution of the work.

During the input inspection the galvanizer shall define a procedure of removing possible defects that hinder proper completion of the work in agreement with the purchaser.

7.2 Quality management system

Current market requirements impose the obligation on the surface treatment contractor to guarantee certain quality assurance conditions. This requirement is met by plants that have quality management system certification in accordance with the respective harmonized standards (EN ISO 9001). The quality management (QM) system requires the contractor to only work with raw materials verified in accordance with approved procedures and to document all production procedures and processes to ensure traceability of the conditions under which a particular work was done and the possibility to analyze the state of the article if necessary. The purchaser is entitled to carry out audits of fulfillment of the agreed conditions of implementation of the work on the contractor's premises.

7.3 Acceptance inspections on order completion

The contractor (galvanizer) is obliged to hand the work to the purchaser for the work to be considered completed. Acceptance inspections are normally undertaken in the galvanizer's plant with the agreement of the purchaser's representative. The object of the inspections is to compare conformity of the galvanized article to the quality standard as defined in the technical conditions defined by the respective technical standard or a special agreement concluded between the galvanizer and the purchaser. In the case of unacceptable defects found during the acceptance inspection of a galvanized article, a decision shall be taken on the responsibility for the occurrence and the process of their remedy. An inspection for quantifiable assessment of quality of the completed work comprises a coating thickness measurement and possibly a measurement of its adhesion if specified.

If a duplex anti-corrosion system is applied (see Section 8.3), the acceptance inspection in the galvanizing plant should also be carried out in the presence of a representative of the organic coating contractor so that the results of the inspection can be used as the input inspection of the paint shop.

7.4 Coating thickness measurement and assessment

The performance and service life of zinc coatings, i.e., also hot-dip galvanized coatings in accordance with EN ISO 1461, are proportional to the thickness of the applied layer. Therefore the thickness of the coating produced by hot-dip galvanizing is one of the material aspects for assessment of the metal plating quality.

Within the batch hot-dip galvanizing process, the coating thickness is difficult to control. It varies in the range between several tens and hundreds of micrometers. This depends on the thickness of the article, its chemical composition, and other conditions as well as the galvanizing conditions.

A proper procedure for measuring coating thickness and observance of the prescribed principles of assessment of measured values is a precondition for objective assessment of galvanizing quality and the reproducibility of processes and results.

7.4.1 Gravimetric method

The reference method for coating thickness determination in case of doubt is the gravimetric method in accordance with EN ISO 1460, which is based on the principle of determining the weight difference of a cut-off piece with a known coated surface area before removal of the coating (by pickling) and after it. The known specific weight of the coating, area of the sample surface, and measured weight of the coating removed by pickling can be used to determine the specific weight of the removed coating layer, which is used to calculate the average layer thickness on the tested surface.

7.4.2 Microscopy method

The microscopy method (Figure 7.1) in accordance with EN ISO 1463, when the coating thickness on a metallographic sample is measured with the use of a microscope scale, is a very suitable method of determining the hot-dip galvanized coating thickness. Its advantage is the possibility of evaluating the assessed coating from the point of view of its structure as well. A disadvantage consists in high time demands and the required equipment as well as in the fact that it is a destructive test. The test sample must be cut off the galvanized part.

7.4.3 Coulometric method

The coulometric method in accordance with EN ISO 2177 is based on the principle of measuring the size of the electric charge necessary to dissolve the coating on an area delimited by an O-ring, which is converted to the volume of dissolved metal. Similarly to the gravimetric method, the thickness is then calculated from the weight of dissolved coating. This is a conditionally destructive method. The coating may be tested without the necessity to cut off the sample; it should be tested in place without the zinc coating and must be repaired in a suitable way in accordance with EN ISO 1461 after the completion of the test.

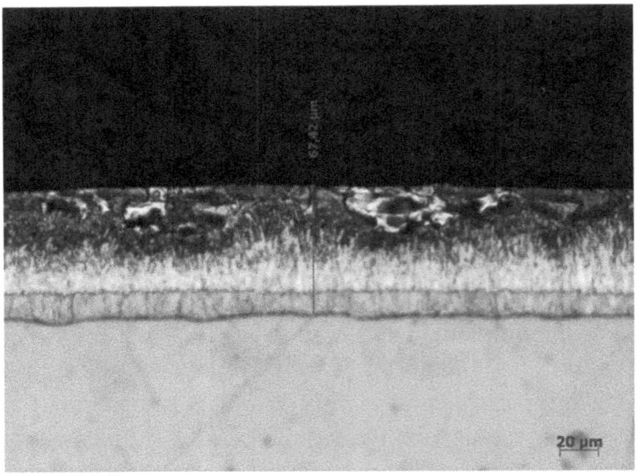

Figure 7.1 Determining the coating thickness using the microscopy method.

Note

EN ISO 1461:2009 does not allow use of the Coulometric method.

7.4.4 Magnetic adherence method

The most frequently used zinc coating thickness measurement methods are magnetic methods in accordance with the ISO 2178 technical standard. They are used to measure the thickness of a diamagnetic material layer on a ferromagnetic metallic base.

The magnetic adherence method is based on measuring the portative force between a permanent magnet and a ferromagnetic base metal (Figure 7.2) that are mutually separated with a diamagnetic layer. The portative force drops with the increasing thickness of this layer. The adherence method only has an informative function. It does not provide sufficiently accurate data for proper evaluation of the measured values.

7.4.5 Measuring the coating thickness using the electromagnetic method

The principle of the electromagnetic method is based on measuring reluctance (magnetic resistance) during the passage of magnetic induction flux through the coating and the basis metal. A sensor of the measuring device excites an electromagnetic field with a certain magnetic flux density. Magnetic induction lines pass through the coating into the ferromagnetic base where they become curved to close in the sensor. Depending on the magnetic resistance of the diamagnetic material of the coating, the density of electromagnetic force lines returning to the sensor is reduced. This reduction is proportional to the magnetic resistance caused by the layer of the diamagnetic material of the zinc coating. The measuring device

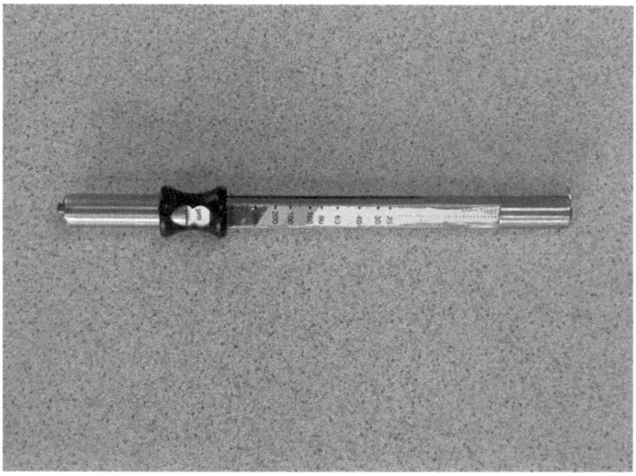

Figure 7.2 Adherence thickness gage.

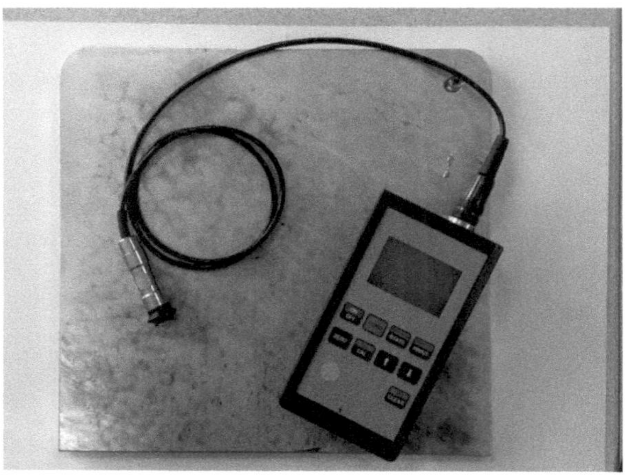

Figure 7.3 Electromagnetic thickness measuring device.

evaluates the difference between the excited magnetic field and the density of the returning electromagnetic force lines and it directly shows the thickness values of the diamagnetic layer in micrometers on the display. A significant advantage of this method is its flexibility (the coating thickness values are determined quickly and immediately during the measurement) and the fact that it is a non-destructive method. However, this method cannot be applied in cases when the basis material is paramagnetic austenitic steel.

Although the thickness values can be read on the thickness gage display with 3 to 4 significant digits (Figure 7.3), the entire measurement is loaded by some uncertainty rate similarly to other methods. You need to know individual factors that may influence the results and eliminate them if possible or take them into account within the evaluation. To make the results of thickness measurement with the use of different methods comparable, the electromagnetic method measurement and its assessment must be carried out in line with the procedure defined in the EN ISO 1461 standard. Rules defined in Chapters 5 and 6 of this standard apply to the zinc coating thickness measurement. Strict observance of the defined rules is a precondition for reproducibility of results.

7.4.6 Factors influencing the coating thickness measurement accuracy using the magnetic method

- **Thickness of the measured coating layer**
  A higher accuracy is achieved with thin coatings where there is a rule that the measurement accuracy does not depend on the wall thickness of the galvanized article. However, in the case of thick coatings the measurement accuracy is influenced by the wall thickness of the base material.
- **Wall thickness of the galvanized part**
  For every device the critical wall thickness of the galvanized part is defined. A measurement performed on an object with a smaller than critical wall thickness is loaded by a significant error.
- **Magnetic properties of the substrate**
  Magnetic properties of different steel grades are different. They mainly depend on the chemical composition of the steel and its structure. The magnetic properties of steel are also influenced by its temperature.
- **Shape and dimensions of the part**
  The method is sensitive to abrupt changes in the shape and to the curvature of the measured surface. Tests carried out within approximately 20 mm from edges or inside a corner are subject to an increased rate of measurement uncertainty. Also, with a decreasing curvature radius of the measured surface the measurement uncertainty rate rises unless the device has been specially calibrated for such tests.
- **Surface roughness**
  On rough surfaces, individual measured coating thickness values generally exhibit a higher spread.
- **Steel texture**
  The magnetic properties of steel are influenced by the direction and method of its machining. Individual measured coating thickness values may differ depending on orientation of the sensor.
- **Remanent magnetism**
  The influence of remanent magnetism is especially significant in the method using a permanent magnet.
- **Foreign magnetic fields**
  The method using a permanent magnet is especially sensitive to the influence of strong magnetic fields generated by various electric devices.
- **Geomagnetic fields**
  When a permanent magnet is used, the measurement results may be influenced by the direction of action of the attractive force of the magnet in the earth's attraction field. For overhead measurements the established coating thickness values are loaded by an error derived from the own weight of the movable part inside the measuring device.

- **Contamination on the measured surface or on the sensor**
  The measurement accuracy depends on achieving a tight physical contact of the sensor with the tested surface. You must check not only the tested surface but also the sensor for cleanness.

7.4.7 *Device calibration*

All measuring devices using the reluctance (magnetic resistance) principle must be calibrated regularly. The calibration is done by the user, who uses a suitable standard according to the manufacturer's operating instruction or externally in authorized calibration facilities. Device calibration must be verified before every measurement, namely in the place of the test when the device is being put in operation.

7.4.8 *Terminology*

EN ISO 1461 defines the terminology used in Chapter 3. If the evaluation of zinc coating thickness measurement is done in accordance with this standard, the terminology must also comply with this.

- **Significant surface** is an area on the product for which the coating is essential from the serviceability or appearance point of view.
- **Control sample** is a group of the defined number of articles selected randomly from the inspection lot that will be subject to testing. Although not defined in the standard, in practice, when more than one section thickness is present within an inspection lot, it is sensible to select articles that are representative of the range of section thicknesses.
- **Inspection lot** is a single production load or order handed over by the purchaser to the galvanizer.
- **Reference area** is a selected area on the significant surface of an inspected article with a size of at least 10 cm^2 and where at least five individual coating thickness measurements must be carried out in each case. In the case of small parts with a significant surface smaller than 10 cm^2, a sufficient number of products will be selected for each reference area so that the total area of their significant surfaces can amount to at least 10 cm^2.
- **Local coating thickness** is the mean value of individual measured values of coating thickness in one reference area.
- **Mean coating thickness** is the average value of local coating thicknesses, i.e.,
  - For each article in the control sample when its significant surface area is greater than 2 m^2
  - For all reference areas in the control sample when their significant surface areas are equal to or less than 2 m^2
- **Minimum value of the coating thickness** is the lowest single measurement in a gravimetric test or the lowest mean obtained from the specified number of measurements in a magnetic test within a reference area.

7.4.9 *Control sample*

The measurement is carried out on a control sample selected in accordance with Table 7.1. The minimum number of randomly selected articles in each control sample depends on the number of articles in the inspection lot. What is decisive for each inspection lot is the actual frequency of occurrence of the section thickness of the galvanized part within the respective interval defined in Table 7.2 or Table 7.3. EN ISO 1461 states that where articles include different thicknesses of steel, each thickness range shall be regarded as a separate article and the relevant required values of coating thickness shall apply.

Table 7.1 Control sample size depending on the inspection lot size in accordance with EN ISO 1461

Number of articles in the lot	Minimum number of articles in the control sample
1 to 3	all
4 to 500	3
501 to 1200	5
1201 to 3200	8
3201 to 10,000	13
over 10,000	20

Table 7.2 Minimum coating thickness on non-centrifuged samples in accordance with EN ISO 1461

Product and its thickness (t)	Local coating thickness	Mean coating thickness
(mm)	(μm)	(μm)
$t > 6$	70	85
$3 < t \leq 6$	55	70
$1.5 \leq t \leq 3$	45	55
$t < 1.5$	35	45

Table 7.3 Minimum coating thickness on centrifuged samples in accordance with EN ISO 1461

Product and its thickness (t)		Local coating thickness	Mean coating thickness
(mm)		(μm)	(μm)
threaded products	$t > 6$	40	50
	$6 \leq t$	20	25
others	$t \geq 3$	45	55
	$t < 3$	35	45

In practice, a suitable approach to differing section thicknesses within an inspection lot and the selection of control samples could be:

- **one control sample** – if the section thicknesses of all parts of that individual article in the inspection lot are made within one interval in accordance with the first column of Table 7.2;
- **two control samples** – if the section thicknesses of all parts that individual articles in the inspection lot are made of are within two different intervals in accordance with the first column of Table 7.2;
- **three control samples** – if the section thicknesses of all parts that individual articles in the inspection lot are made of are within three different intervals in accordance with the first column of Table 7.2;
- **four control samples** – if the section thicknesses of all parts that individual articles in the inspection lot are made of are within four different intervals in accordance with the first column of Table 7.2.

Example (see Figure 7.4):

An inspection lot contains 1500 weldments, all of them being made of a structural material with a section thickness exceeding 6 mm; 750 parts of the same inspection lot also

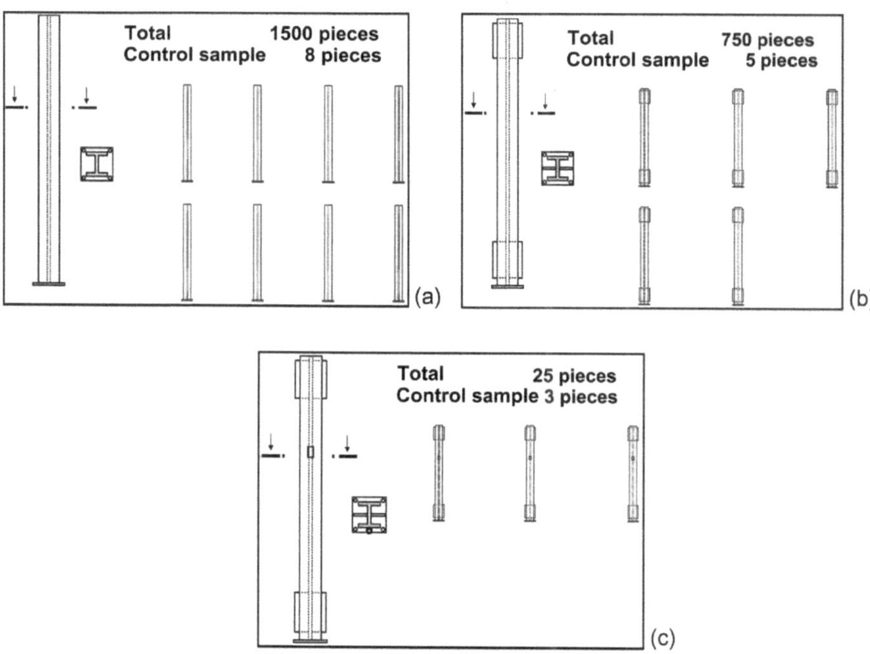

Figure 7.4 Production lot of 1500 pieces composed of various HEB 120 posts – a half of the production lot is made with the use of sheet-metal figures with the thickness of 4 mm and 25 pieces contain tube bushings with Ø 30 × 2 mm: a) control sample for the HEB 120 post, b) control sample for sheet-metal figures with the thickness of 4 mm, c) control sample for tube bushings with Ø 30 × 2 mm.

contain other elements of a different structural material with a section thickness of 3 to 6 mm, and, in addition, 300 parts of this inspection lot contain applications made of a structural material with a section thickness smaller than 3 mm.
In this case, three control samples will be selected for the inspection measurements out of which the first one, for coating thickness measurement on a substrate thicker than 6 mm, contains at least eight products; the second one, for coating thickness measurement on a 3 to 6 mm substrate, contains at least five products; and the third control sample, for coating thickness measurement on a substrate with a lower section thickness than 3 mm, consists of at least 3 products. This example may exceed the minimum requirements of EN ISO 1461 but can be considered good practice when dealing with a mix of section thicknesses within an inspection lot.

7.4.10 Reference area

A reference area must lie inside an area suitable for application of the gravimetric method, i.e., inside a place whose area is easy to determine.

The number of reference areas on individual products in the control sample depends on their surface and mainly on the area of their significant surface. On long products, reference areas should be selected about 10 cm from each end and also approximately in the middle of their length. Each reference area must comprise the complete cross-section of the product (see Figure 7.5).

- At least three reference areas per product should be selected for large article with the significant surface larger than 2 m^2.
- For products having their significant area larger than 100 cm^2 but less than 2 m^2, at least one reference area must be selected on each article in the control sample.
- For products having their significant area between 10 cm^2 and 100 cm^2 at the most, just one reference area must be selected on each article in the control sample.
- In the case of small parts with a smaller significant surface than 10 cm^2, a sufficient number of articles shall be selected for each reference area without repeating so that the total area of their significant surface can amount to at least 10 cm^2. If the number of products in the inspection lot is smaller than the sum of articles in all groups that need to be created to obtain the prescribed number of sufficiently sized reference areas, then the number of articles in the control sample is equal to the number of articles in the inspection lot. The number of reference areas of such a control sample is equal to the number of article groups selected for individual reference areas.

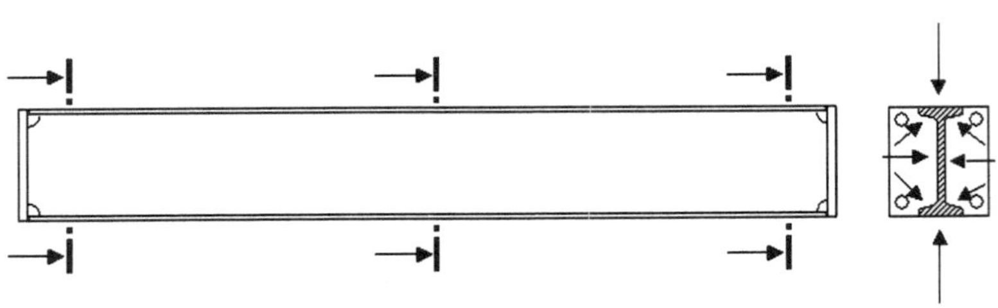

Figure 7.5 Reference areas on long products.

7.4.11 Coating thickness

In each reference area, whose area must not be smaller than 10 cm^2, at least five individual coating thickness values must be measured. Then, the local coating thickness of each reference area is the arithmetic mean of these individually measured values. The coating thickness must not be measured in places that are located less than 10 mm from discontinuities (edges, holes, inside corners) or flame-cut surfaces. Thickness measurement should be avoided on strongly curved surfaces unless the measuring device has been specially calibrated for such measurements.

Individual measured coating thickness values that are subsequently used to calculate the local coating thickness are not significant for the assessment. It is only their arithmetic mean that is assessed without the condition of not exceeding the limit in accordance with Table 7.2 or Table 7.3 having to be met by each coating thickness value established by measurement.

7.4.12 Assessment of measurement results

For proper assessment of the established coating thickness value individual, checked articles must be correctly classified. Here, a decision must be made about two essential parameters that characterize each product:

- **Section thickness of the galvanized part**
  If an article in the inspection lot is made of structural material whose section thicknesses belong to more intervals defined in Table 7.2 or Table 7.3, such an article must be assessed within all control samples of the particular inspection lot where it is represented based on the section thickness interval. A separate measurement and coating thickness assessment must be carried out for each control sample.
- **Significant surface area**
  If the area of the significant surface of an inspected product is larger than 2 m^2, then the mean coating thickness value determined for a product with a functional surface area larger than 2 m^2 is only valid for the particular article. The mean coating thickness value of products in a control sample whose significant surface area is 2 m^2 at the most is calculated as the arithmetic mean of all local coating thicknesses in the whole control sample. If one control sample contains products with a significant surface area up to 2 m^2 as well as over 2 m^2, the record of inspected articles from the same control sample shall contain different mean coating thickness values calculated in accordance with the above mentioned condition.

The assessment of performed measurements must contain the results of a comparison of each calculated local coating thickness and each mean coating thickness to the minimum values prescribed by the standard in Table 7.2 or Table 7.3. The result of this comparison should be summarized in the final statement clearly declaring whether the zinc coating thickness of the particular inspection lot meets or does not meet the requirements imposed by the EN ISO 1461 standard.

Note

The fact that the mean coating thickness of products with a significant surface area of 2 m^2 at the most is calculated and assessed for the entire control sample makes it possible to even classify such a product that would not meet the prescribed conditions separately as acceptable.

7.5 Coating adhesion tests

Adhesion of galvanized coating to the substrate does not usually need to be tested. Unlike organic coating materials, where the performance and durability of anti-corrosion protection is directly proportional to the adhesion of the protective layer to the substrate and its consistency, the anti-corrosion effect of a zinc coating is mainly related to cathodic protection. It only becomes efficient when both the electrically conductive metals are in contact, regardless of possible cracks and pores in the coating (see Section 8.1). A properly applied galvanized coating on steel that is suitable for hot-dip galvanizing usually exhibits sufficient adhesion to the substrate. However, what should be kept in mind is that parts with a thick coating require especially careful handling. Mechanical processing of hot-dip galvanized products is considered as non-standard handling of the product from the point of view of coating adhesion.

If the surface treatment purchaser requests adhesion tests, such a test and the required parameters may be agreed with the galvanizer on placing of the order. Adhesion can be measured with the use of a pull-off test in accordance with EN ISO 4624 or EN ISO 16276-1 by means of a special fixture measuring the force required to pull off the coating on a delimited area (Figure 7.6). A suitable adhesive

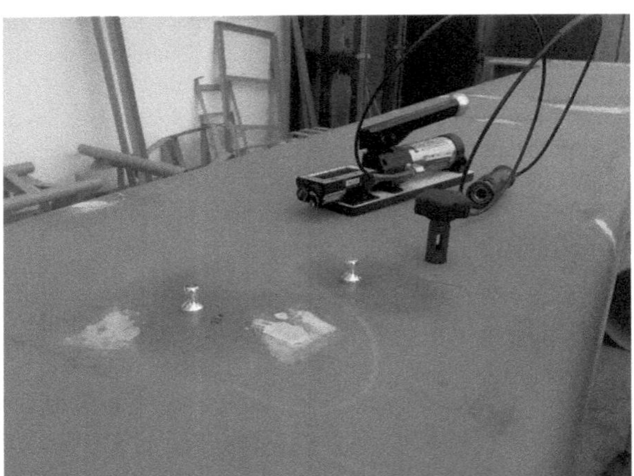

Figure 7.6 Pull-off adhesion tester (two bonded dollies in the foreground).

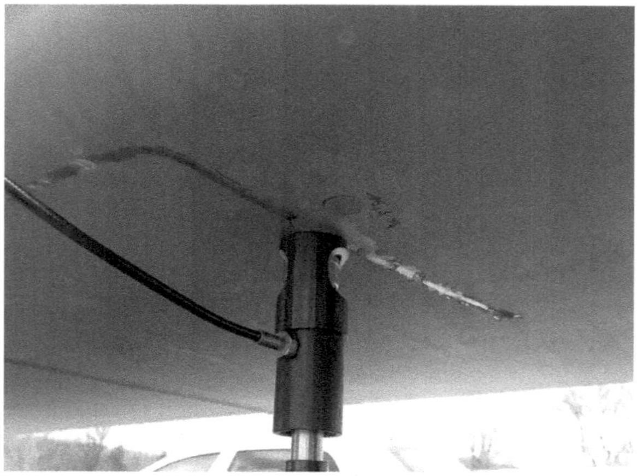

Figure 7.7 Probe acting upon the dolly.

is used to bond a test dolly of corrosion-resistant steel onto a flat surface of a galvanized part. After hardening the adhesive a circular target is marked around the contact surface of the dolly using a fixture and the dolly is pulled off using a special device (pull-off adhesion tester) (Figure 7.7) while the tension required to pull off the dolly is measured. The boundary between the coating layers (or in the adhesive) where the dolly was pulled off is also assessed.

7.6 Zinc corrosion

Zinc is a metal that exhibits relatively low corrosion resistance in its pure form. If pure zinc is exposed to atmospheric influences, it oxidizes quickly. Depending on the prevailing conditions, the following reactions are the most frequent ones:

$$2Zn + O_2 \rightarrow 2ZnO$$

(zinc oxide is a white substance soluble in diluted acids)

$$2Zn + 2H_2O + O_2 \rightarrow 2Zn(OH)_2$$

(zinc hydroxide is a white substance with poor solubility in water)

$$Zn(OH)_2 + 2CO_2 \rightarrow Zn(HCO_3)_2$$

(zinc hydrogen carbonate is a white substance soluble in water)

$$ZnO + H_2SO_3 \rightarrow ZnSO_4 + H_2$$

(zinc sulphate is a white substance soluble in water)

$$ZnO + 2HCl \rightarrow ZnCl_2 + H_2O$$

(zinc chloride is a white substance soluble in water).

The high durability of the anti-corrosion protection by galvanized coating results from its ability to be passivated (see Chapter 8). In favorable conditions (in a dry, ventilated environment) atmospheric humidity causes the gradual transformation of zinc oxide to zinc hydroxide, which can react with the atmospheric carbon dioxide and change to the relatively stable zinc carbonate. Each chemical reaction runs at a certain rate under the given conditions. If the zinc hydroxide formation rate is equal to its transformation rate to zinc carbonate, the zinc coating gets covered by patina. However, if in the period before formation of a sufficient patina layer the galvanized part is wetted, the volume of the produced zinc hydroxide will exceed the formation rate of zinc carbonate and the coating will be covered by very voluminous corrosion products of zinc. Corrosion attack of zinc is substantially supported by high air pollution, especially an increased concentration of sulfur dioxide or chlorides in the atmosphere. The zinc coating will be covered by corrosion products, generally white, hence the name white rust (Figure 7.8). However, corrosion products of zinc may have various color shades depending on their origin [45].

Figure 7.9 shows a zinc coating facing a corrosive attack by extract from a wooden balk used to support a galvanized part. This is a superficial defect that does not impair the corrosion resistance of the entire system, and the coating loss until the stain disappears due to patina formation is negligible.

The causes of occurrence of white rust are not related to the quality of the hot-dip galvanized coating. This defect must be understood as a phenomenon that is inherent

Figure 7.8 White rust on a zinc coating caused by chlorides contained in road salt.

Figure 7.9 Zinc coating affected by aggressive extract from a wooden balk used to support galvanized articles during transport or storage.

Figure 7.10 Galvanized supports after a yearly unsuitable exposure, when they were partly immersed in a dam reservoir.

to the principle of the applied anti-corrosion system and the way of handling of the galvanized part. Therefore, white rust cannot be generally a reason for rejection.

A galvanized coating provides steel parts with reliable and long-term protection from corrosion if they are exposed in a suitable environment. Galvanized parts perform best when exposed to climatic influences in locations with low air pollution and where wetting of their surface is only temporary. If the corrosion conditions are unsuitable for hot-dip galvanized coating (Figure 7.10) the service life of the coating may be very short under certain conditions (see also Chapter 8). The decision to select an anti-corrosion system for a particular environment must be justified by a prior evaluation of its suitability.

7.7 Coating defects and verification of an attainable quality standard

High performance and long-term durability of an anti-corrosion system based on hot-dip galvanizing are not very sensitive to the quality of the applied coating. In spite of this, there are certain quality standards that govern acceptance inspections. Substantial features are the thickness of the coating and its possible absence on the significant surface. Coating thickness is the only compulsory characteristic that can be accurately quantified within the assessment procedures prescribed in the EN ISO 1461 standard. In case of an absence of coating the rule is that all places without galvanized coating on a galvanized part must be repaired. In this case the sum of bare areas designed to be repaired must not be larger than 0.5% of the total area of the part surface and at the same time an individual uncoated area to be repaired must not be larger than 10 cm^2. However, the purchaser and the galvanizer may conclude a different agreement.

The principle valid for batch hot-dip galvanized coatings is that roughness is a relative parameter and it is evaluated mainly from the point of view of functionality of the respective surface on the part. If the purchaser puts higher emphasis on a smooth surface, the attainable quality standard should be verified in advance by use of a sample. Usual coating appearance tests are conducted from a distance of up to 1 m by visual inspection without optical aids. There must not be any sharp protrusions on the coating that might cause an injury. Vacant places on surfaces where the coating is important in terms of their function are unacceptable.

The occurrence of darker or lighter places or some uneven places on the surface as well as the presence of white rust must not constitute a reason for rejection to accept an order if the acceptance inspection proves that the coating thickness is higher than the prescribed minimum value.

Adhesion of the coating (see also Sections 7.5 and 7.8) to the substrate is not commonly tested as it depends on the properties of the substrate material. Commercial galvanizing plants galvanize non-standard parts made of various materials, often unspecified and mutually combined substrate materials in the case of weldments. Setting specific parameters for coating applications of every material separately is impossible. The conditions for batch hot-dip galvanizing are optimized in such a way to make its characteristics on commonly used substrates (structural steel) as uniform as possible.

7.7.1 Laps and flakes

Some rolled semi-products are affected by surface defects as laps (Figure 7.11) and flakes. These defects are usually filled by scale. During pickling, wüstite present in the gaps under the laps is dissolved and the pickling acid or flux penetrates into the resulting cavities. Subsequently, this solution, saturated with iron salts, boils over during immersion of the part in the zinc bath and contaminates the surrounding part surface. Contamination of decontaminated surface by boiling chlorides near laps causes coating defects (Figure 7.12).

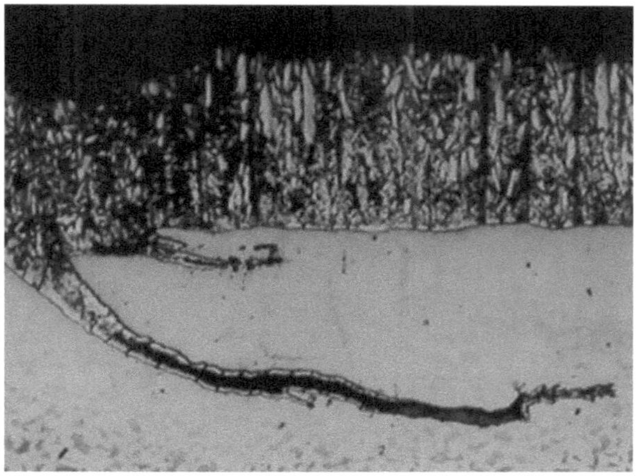

Figure 7.11 Lap on substrate steel.

Figure 7.12 Surface defects caused by laps.

7.7.2 Insufficient pickling

A precondition for creation of iron-zinc alloy coating is clean metallic substrate. If the pickling time is not sufficient, residues of oxides and scale on the part surface prevent metallurgical reaction between iron and zinc and cause defects called black spots (Figure 7.13).

Modern production processes in rolling mills comprise the operation of scale removal from rolled material. If a part to be hot-dip galvanized is made of scaled material, the presence of hematite on the steel surface may not be discovered in the galvanizing plant. Ferric oxide is very poorly soluble in hydrochloric acid and its continuous layer may persist on a substantial part of the galvanized article even after pickling. Iron in the form of oxides cannot be released in the zinc bath and

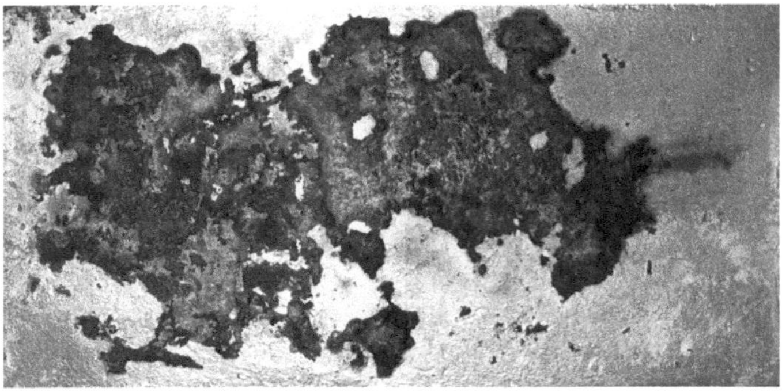

Figure 7.13 Black spot caused by insufficient pickling.

Figure 7.14 Metallography of coating on scale.

therefore it does not participate in the metallurgical reaction with zinc. A scaled surface of steel activated by the applied flux may be sufficiently wettable in molten zinc under certain conditions and pure zinc adheres to such a substrate (Figure 7.14). After emerging from the zinc bath it solidifies and in such a case the zinc coating exclusively consists of pure zinc (phase η), and no iron-zinc alloy phases are formed (see also Section 7.8.5). Unlike the alloy coating the adhering layer of pure zinc is not interconnected with the substrate and its thickness and adhesion are very low. Such coating is, besides easy delamination, characterized by the good malleability of a peeled layer (Figure 7.15).

7.7.3 Superficially contaminated substrate

A cause of pickling inefficiency, in spite of its sufficient duration, may be that an impurity on the substrate surface cannot be removed in the usual way during the degreasing operation (Figure 7.16). An insoluble film prevents perfect wetting of the part to be galvanized in the pickling agent and the substrate remains contaminated by impurities that prevent wetting of the steel surface in liquid zinc and are the cause of uncoated areas, i.e., the "black spots" (Figures 7.17 and 7.18).

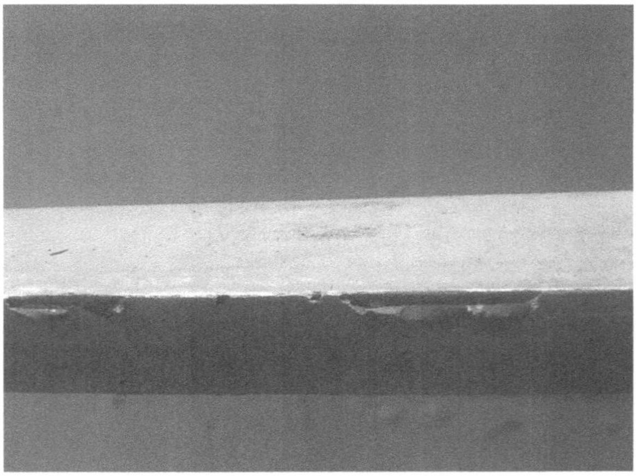

Figure 7.15 Malleable pure zinc coating on scaled surface.

Figure 7.16 Superficially contaminated substrate.

Figure 7.17 Black spots on a substrate contaminated by a poorly soluble substance.

Figure 7.18 Places without coating caused by the use of an unsuitable separation agent for welding.

Note

Even separation agents that are declared to be suitable for hot-dip galvanized structures by the manufacturer must be applied adequately. Their overdosing may also lead to galvanizing defects.

7.7.4 Jigging marks

For hot-dip galvanizing of heterogeneous non-standard parts with the use of suspension jigs, these parts must be suitably attached to the jigs so that they can be subsequently handled in a mechanized way. In practice, various fixtures, chains, and mainly wires are used to suspend the parts.

In the place of mutual contact the galvanized part gets soldered to the jig and subsequently, when the part is removed from the jig, the zinc coating is peeled off in this place (Figure 7.19). The edges of the contact place are affected by burrs. Beams are usually ended with junction plates with holes that can be advantageously used to hang for galvanizing. Although by using suspension devices (Figure 7.20) or the chain (Figure 9.2), each area without zinc coating is not greater than allowable area 10 cm^2, yet the galvanizer should require an approval of such way of hanging by the customer because depreciation (Figure 7.21) of the part may be undesirable.

7.7.5 Marking of components and fabrications

Parts handed over for hot-dip galvanizing must not be marked with any pigment or markers that are insoluble in water. This is because no galvanized coating will be generated on surface marked like this (Figures 7.22 and 7.23).

Figure 7.19 Contact place after a wire.

Figure 7.20 Use of the suspension devices.

Figure 7.21 Contact place after a fixture for suspending a part during hot-dip galvanizing.

Figure 7.22 Marking of a part with organic pigment.

Figure 7.23 Defect caused by unsuitable marking of a part.

7.7.6 Assembly welds

Sometimes, hot-dip galvanized parts must be welded to form larger units during the assembly. Although the EN ISO 1461 does not accept the possibility of repairing a spot with missing coating on a larger area than 10 cm^2, this provision is valid as a condition for acceptance inspections between the galvanizer and the purchaser. It does not preclude the possibility of concluding another agreement or installation of welded joints or other additional modifications of hot-dip galvanized parts by the purchaser. The EN ISO 14713-1 standard sets out that the zinc coating should be removed from the weld surfaces and their vicinity to ensure the best quality of the welded joint and to prevent the welder's exposure to breathing of zinc vapor, which can cause a temporary "zinc fever."

Figure 7.24 Assembly weld of a galvanized structure made on-site.

Surfaces without zinc coating in the place of the weld must be suitably additionally repaired to equalize the efficiency of anti-corrosion protection with the surrounding zinc coating (Figure 7.24).

Note

Given the extent of repairs needed for assembly welds preferable to use is the technique of thermal spraying (see also Section 7.9.1).

7.8 Delamination of a hot-dip galvanized coating

The assessment of conformity is based on the technical conditions set out in EN ISO 1461. They mainly focus on the thickness of the applied coating and the required function of the galvanized surface with a view to possible risk of injury by sharp protrusions or lumps. One of the possible quality assessment aspects is coating adhesion. However, here the standard does not specify any particular conditions and leaves their specification up to the agreement between the purchaser and galvanizer.

In the case of organic coating, adhesion to the substrate is one of the decisive factors influencing performance and useful life of the surface finish. Adhesion does not have a direct impact on efficient zinc coating protection, as cathodic protection is efficient at mere contact of zinc with steel. A precondition is their mutual conductive bonding in a galvanic cell. Hot dip galvanized coatings are expected to have sufficient adhesion to be able to withstand common external influences. However, insufficient coating adhesion may represent a major obstacle for a particular application.

Problems with insufficiently adhesive galvanized coating occur when the galvanized part is exposed to external influences. A coating with reduced adhesion is not able to resist even moderate mechanical load. This holds good, e.g., for sweep blasting carried out for preparation of the surface under organic coating material in duplex system, which may lead to surface damage.

Delamination of galvanized coatings has conditional significance for the performance and useful life of the galvanized anti-corrosion protection to a considerable extent. In normal practice, local surface delamination may not be manifested in any way during its useful life since even coating affected by local detachment is able to protect the treated part from corrosion efficiently and for the long term.

7.8.1 High silicon content

Silicon contained in "killed steel" causes increased reactivity of iron with zinc. On steel with silicon content above approximately 0.28% galvanized coatings are relatively thick and this phenomenon is even multiplied by a high thickness of the substrate steel section. In extreme cases the coating thickness approximates 1 mm (Figure 7.25). In the structure of such coating the predominant phase is phase ζ, which consists of compact monoclinic crystals characterized by a relatively high degree of randomness. Between individual crystals of phase ζ there are frequent discontinuities; the bond of the coating to the surface is not established over the whole surface, but only at certain points (Figure 7.26).

Due to different thermal expansion coefficient values of the coating and steel, considerable tangential stress already occurs between the coating and the substrate in the cooling stage of galvanized parts, which may cause local spontaneous delamination of the coating. Galvanized products made of steel with high silicon content are not suitable for application of a duplex system. Sweep blasting damages the galvanized coating (Figure 7.27).

In an extreme case, on a very reactive substrate with a very high silicon content and very thick wall, the coating may grow to an extreme thickness in the

Figure 7.25 Coating delamination on steel with high Si content.

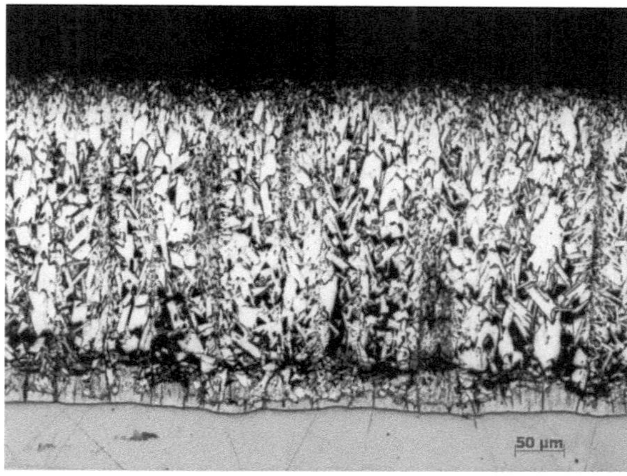

Figure 7.26 Metallography of coating on steel with high Si content.

Figure 7.27 Coating on steel with high Si content after sweep blasting.

range of a few millimeters. Volume changes during an intensive metallurgical reaction in the course of the dwell of the galvanized part in the zinc bath cause collapses of the coating already in the stage of its formation. Within the process the substrate is exposed and very thick coating is formed again. The galvanized part is covered by an inconsistent crust (Figure 7.28). Parts galvanized this way are useless for their purpose.

7.8.2 Silicon content in the Sandelin range

Steel with silicon content in the Sandelin range (i.e., approximately between 0.03% and 0.12%) is especially unfavorable for hot-dip galvanizing. During

Figure 7.28 Fe-Zn crust on massive elements of steel with very high Si content.

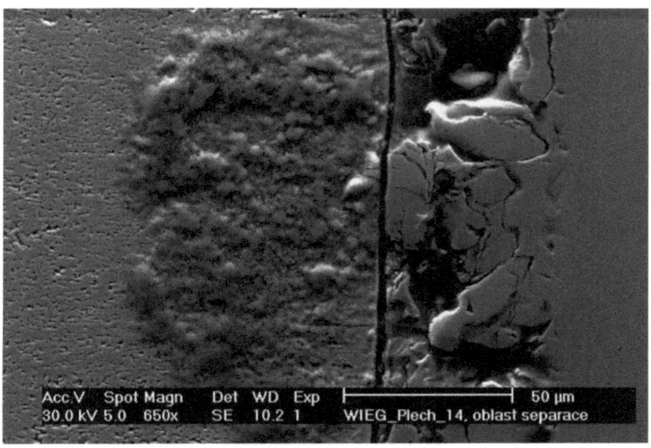

Figure 7.29 Shrinkage porosities in the surface structure of steel caused by diffusion during hot-dip galvanizing.

hot-dip galvanizing, silicon that was substitutionally dispersed in the crystal lattice of steel causes an excessive release of iron atoms. These then strongly diffuse to the zinc melt surrounding the walls of the galvanized part. There, they bond with zinc into the 1:13 ratio, producing very fine crystals of hard zinc (phase ζ), which form a highly viscous, mushy substance together with pure zinc. This substance adheres to the walls of the galvanized part in a thick layer. After solidification the created coating has a relatively very good consistency, but very low adhesion to the surface is caused by shrinkage porosities in the surface layer of the substrate (Figures 7.29 and 7.30).

Figure 7.30 Coating on the surface of thin sheet with Si content of 0.08% affected by shrinkage porosities.

Figure 7.31 Delamination of coating on steel with Si content in the Sandelin range.

Predominant diffusion of iron in a system where steel and the iron-zinc alloy meet at the phase boundary causes formation of shrinkage porosities, Kirkendall vacancies in the steel (see also Section 4.1). The principle of this phenomenon is that the flux of iron atoms from the substrate is balanced by a flux of vacancies to steel. Shrinkage porosities occurring in the steel structure may be so important that they may even disturb its surface layer. The disintegrating surface layer of steel causes a loss of contact of the applied iron-zinc coating with the substrate (Figure 7.31). Co-acting unfavorable influences are tangential stresses at the boundary between the coating and substrate as well as hydrogen escaping from the substrate, which recombines in the produced vacancies and material discontinuities.

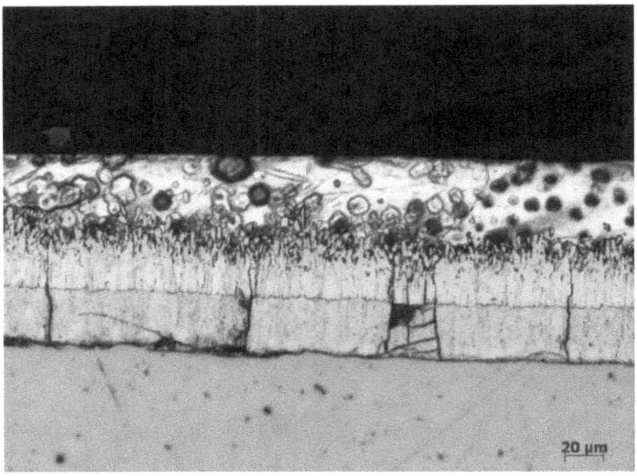

Figure 7.32 Radial cracks in the coating.

7.8.3 Primary delamination

During the dwell of the galvanized part in the zinc bath, when the alloy coating matures, the substrate is heated up, which causes its expansion. This is the cause of possible formation of cracks in the coating. If these cracks are radially oriented (Figure 7.32), their presence in the coating is not important, they do not reduce its adhesion to the substrate, and they do not even reduce the corrosion resistance of the coating as during wetting. The sacrificial anode principle is applied and iron passes into the state of immunity. It does not corrode even if moisture creeps up to the substrate (see also Chapter 8).

In an unfavorable case, the shear stress between the coating and steel may lead to the formation of tangential cracks and cause detachment of the generated coating from the substrate. If zinc melt has an opportunity to penetrate into such a cavity, zinc immediately reacts with the substrate and a new coating layer is formed there. It may crumble away due to the influence of mutual shifts of the top coating and the substrate, and the cavities are filled with incoherent matter (Figure 7.33). Coatings applied onto steel with very low silicon content are very prone to primary delamination. These coatings are consistent, and especially the phase δ layer of close packed hexagonal crystals represents an efficient barrier to escaping hydrogen recombining at the boundary between the substrate and coating. Accumulating molecular hydrogen supports delamination. The surface of the applied zinc coating is rough (Figure 7.34).

7.8.4 Delamination on thermally cut surfaces

At the untreated edge of a thermally cut edge (Figure 7.35) where the natural surface of rolled material passes into the surface caused by steel burning, a step change of material properties occurs. On the thermally cut surface, zinc reacts with the substrate

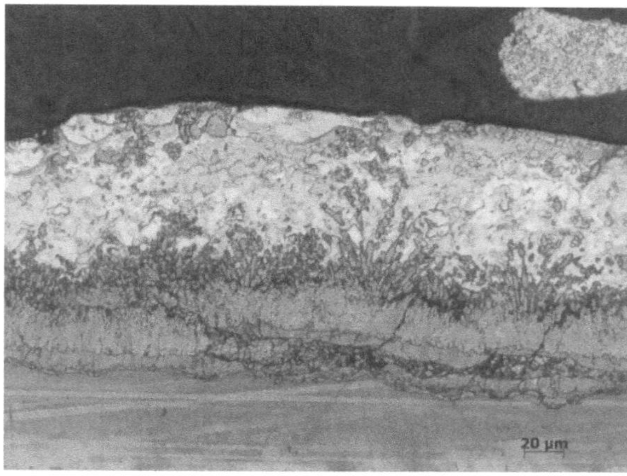

Figure 7.33 Primary delamination in a tangential crack.

Figure 7.34 Rough surface caused by primary delamination.

differently from normal surfaces. Therefore, the properties of the galvanized coating abruptly change here, too. On both the contact surfaces, the produced coating has a different structure and adhesion. Simultaneous action of substantial temperature stress often causes spontaneous delamination of coating on the thermally cut edge (Figure 7.36). The surfaces of thermal cuts should be machined; unchamfered edges represent an increased risk of delamination (see also Section 6.3).

7.8.5 Continuous hematite layer

During the production of steel semi-products their surface is covered by a scale layer (see Section 3.2.1). These are iron oxides, especially ferrous oxide (wüstite FeO) and ferric oxide (hematite Fe_2O_3) or their mixture magnetite (ferro-ferric

Figure 7.35 Thermally cut edge.

Figure 7.36 Zinc coating delamination on the edge of a thermal cut of steel.

oxide Fe_3O_4). The bottom layer consists of wüstite, which is relatively soluble in the pickling solution. It is usually covered by a layer of less soluble magnetite or very poorly soluble hematite. In modern plants, the manufacturers of steel semi-products have equipment for scale removal (with a water jet or by pickling), but some semi-products are sent for further processing with scale.

Galvanizing plants do not have equipment to detect presence of hematite on a steel surface. Chemical pre-treatment to achieve a pure metallic surface is done on the basis of experience. The batch is immersed in the pickling bath depending on its character for 1 to 3 hours and in exceptional cases for more than 8 hours. If the hematite layer is continuous, its dissolution is difficult and usual pickling times are not sufficient to remove the scale (see also Section 7.7.2).

Figure 7.37 Malleable pure zinc stuck to a scale layer.

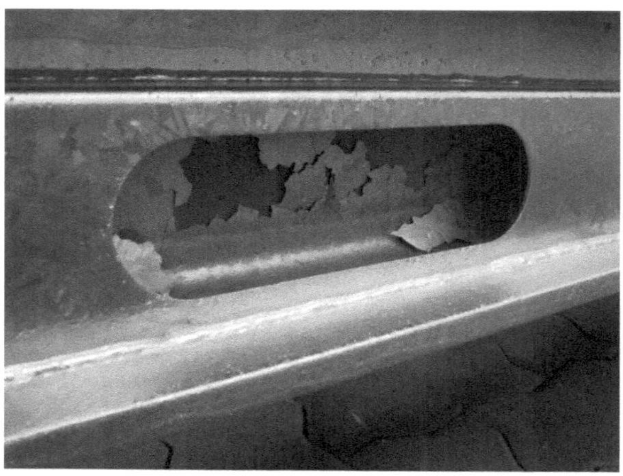

Figure 7.38 Delaminating alloy coating is brittle.

If a scaled surface of steel is activated by flux, it is sufficiently wettable in the zinc melt for zinc to adhere to it. However, the scale layer prevents the metallurgical reaction between iron and zinc as iron has higher affinity to oxygen, with which it is combined in oxides, and at the galvanizing temperature it is not able to be released from them to react with zinc. The fact that the applied zinc coating does not contain iron-zinc alloy phases may not be visually detectable on the integral coating. However, pure zinc does not manifest sufficient adhesion to the substrate and delaminates easily. Coatings consisting of pure zinc are characterized by their relatively thin layer and very good malleability (Figure 7.37) while the delaminating alloy coating is very brittle and prone to cracking (Figure 7.38).

7.8.6 Blisters

The occurrence of blisters is usually accompanied by so-called "mixed structures." Blisters are formed in coatings generally applied to thin substrates with unevenly dispersed silicon in an average concentration near the bottom limit of the Sandelin range, namely at the moment of removal of the galvanized part from the molten metal.

The formation of a blister is caused by hydrogen escaping from the substrate and recombining in numerous discontinuities of the growing coating, where it increases its volume. Blisters are formed over the layer of the generated alloy phases but are covered by a viscous matter consisting of fine crystals of phase ζ that are dispersed in the zinc melt. During its solidification this mushy matter prevents hydrogen from escaping and lens-like cavities are created (Figure 7.39). The surface of the coating gets rough (Figure 7.40). Coating with blisters may be prone to damage during sweep blasting.

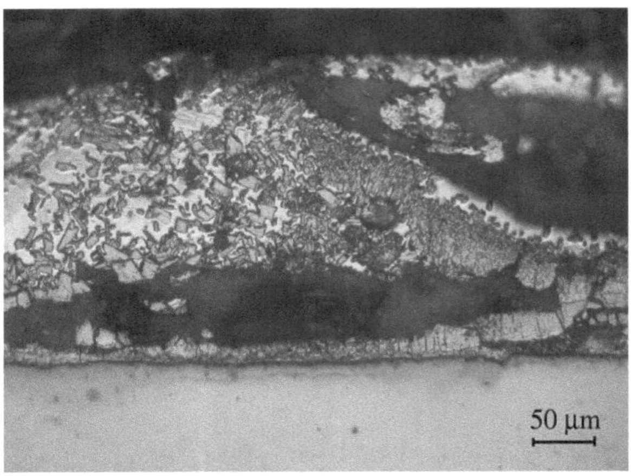

Figure 7.39 Cross-section of a blister.

Figure 7.40 Appearance of coating affected by blistering.

7.8.7 Flat delamination over phase ζ

A successful metallurgical reaction between iron and zinc requires proper activation of the pure metallic part designed to be zinc coated with flux before dipping in molten zinc. Due to flux burning the steel surface does not have a perfect contact with the liquid metal at the moment of its immersion in the zinc bath. The substrate is separated from the liquid metal by a variably thick layer of gas from the burnt flux, which gradually disappears. The viscous mushy matter of fine crystals of hard zinc dispersed in the zinc melt may be the cause of retaining a gas layer in the coating.

Steel with evenly spread silicon in a concentration approximating the bottom limit of the Sandelin range is prone to surface delamination. The mechanism of occurrence of flat delamination on a sheet with the Si content of 0.03% is shown in Figures 7.41 and 7.42. Figure 7.41 shows a gas layer from burnt flux between the

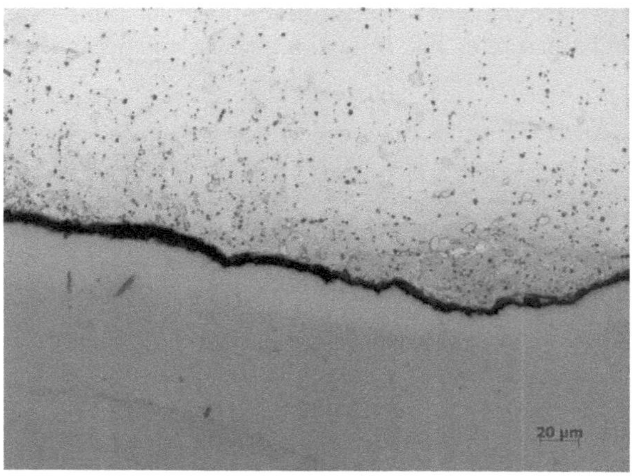

Figure 7.41 Layer of gas from flux burning between the substrate (at the bottom) and the coating metal immediately after immersion of the part in the zinc bath.

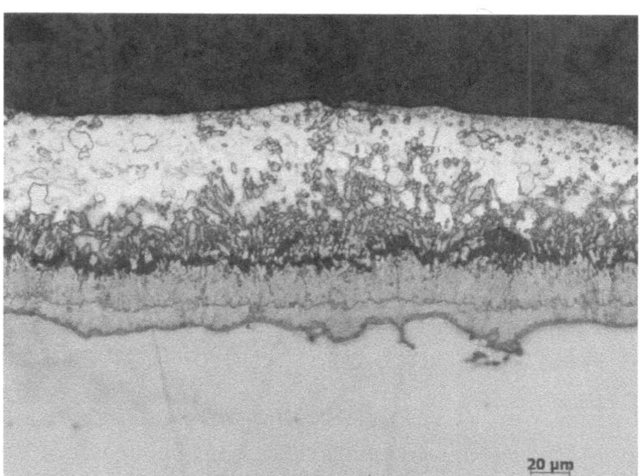

Figure 7.42 Flat discontinuity in the coating over the ζ phase layer.

substrate and liquid metal after 10 s dwell of the sample in the bath. Over the gap in Figure 7.42 there is a clearly visible mixture of small monoclinic crystals of phase ζ overgrown with pure zinc in the coating. In the solid-liquid state the mixture is very viscous and prevents easy escaping of gases.

7.9 Repairs of defects of galvanized coating

During batch hot-dip galvanizing, the galvanized parts are always affected by coating defects in the contact places with suspension methods where soldering occurs and after releasing the coating is peeled off. Even though the cathodic protection works reliably in spite of minor surface defects, all places with absent zinc coating must be repaired in a suitable way. Such places may also be spots with missing coating due to a surface defect of the substrate material, peeled off coating at the edge of a galvanized part, etc. The following principles apply to repairs of coating defects:

- One area on a galvanized part without coating that is to be repaired must not be larger than 10 cm^2 unless another agreement is concluded between the purchaser and galvanizer.
- The sum of all areas without coating on a galvanized part that are to be repaired must not exceed 0.5% of the total surface area of the galvanized part.
- A repair may only be implemented with the use of one of the methods defined in EN ISO 1461.
- The thickness of the repair material layer on the repaired surface must be at least 100 μm.

The EN ISO 1461 technical standard specifies the following possible procedures for repairs of galvanized coatings:

- thermal spraying
- zinc-rich paint containing zinc dust
- coating of zinc flake
- zinc paste
- zinc alloy solder

7.9.1 Thermal spraying

Anti-corrosion protection quality thermal spraying (Figure 7.43) is the most suitable method of repair of defects whose quality, performance, and durability approximate the properties of a dip applied coating. However, it is a costly method with high demands on time and equipment. In most cases the application of this method, similar to zinc solder, is uneconomical.

7.9.2 Repairs with a zinc-rich paint containing zinc dust or zinc flake

The most frequent repair method of coating defects is application of a suitable coating with a high content of zinc dust in the dry film [46]. It is a coating material with a high content of pigment consisting of zinc particles that are bound by a suitable binder.

Figure 7.43 Thermal spraying of zinc.

Figure 7.44 Kit for surface cleaning before application of the repair coating material.

The zinc particles may have a globular or a lamellar shape. Repair coating materials containing lamellar zinc particles (flakes) are more convenient as they are able to create a better barrier at simultaneous mutual long-lasting conductive contact to ensure cathodic protection.

Before the repair it is first necessary to remove burrs, slag, zinc ashes, and other rough contaminants from the affected places using a wire brush, file, or sand paper. The surfaces without a zinc coating prepared this way should be subsequently thoroughly finally cleaned, i.e., wiped, dried, dusted, and degreased using a solvent if necessary (Figure 7.44).

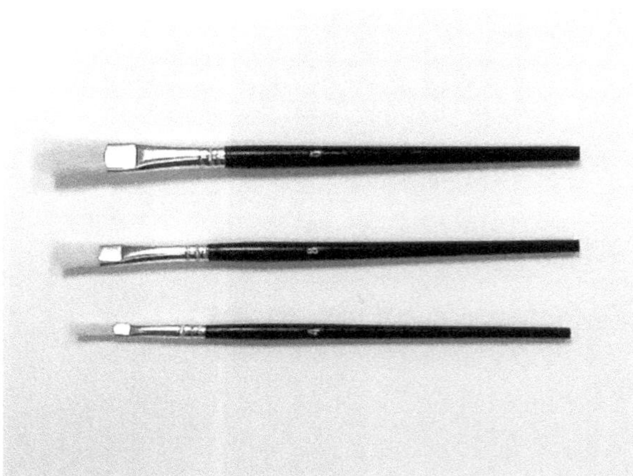

Figure 7.45 Small flat brushes are suitable for application of the repair coating material.

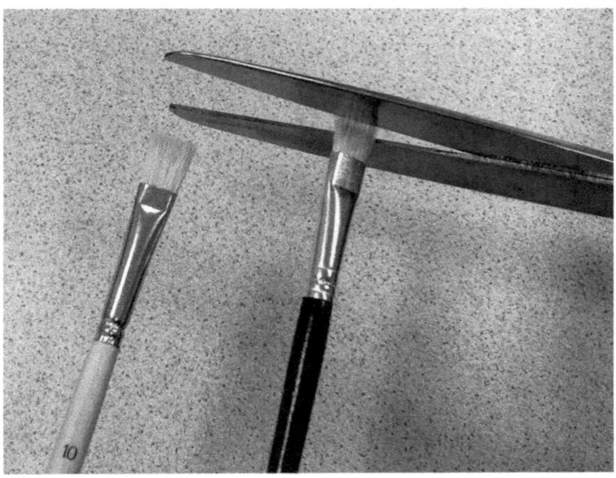

Figure 7.46 Work with a brush with shortened hair is easier.

The repair coating material is applied to damaged or uncoated places without overlapping the surrounding coating in a thickness of at least 100 μm. On small surfaces of the type of scratches, cracks, or peeled edges a sufficiently thick layer of the repair coating can be applied in one layer; on larger surfaces the process must be repeated after about 5 minutes when the first coating layer has dried slightly.

For proper application of the repair coating it is important to select a suitable brush (Figure 7.45). Working with a small flat brush with shortened hair is easier (Figure 7.46). The repair coating is applied in a large layer without spreading

Figure 7.47 A light color shade of the coating material gives a disturbing impression.

Figure 7.48 A dark shade of the repair coating material is not very distinctive on a light coating.

by stippling as if the covered area should be sealed with the coating material. The coating is applied undiluted, well stirred.

The anti-corrosion properties of the coating produced with the repair coating material with a high content of zinc dust are not comparable to the anti-corrosion properties of integral zinc coating in terms of quality. The anti-corrosion protection of the repair coating is better and more efficient the thicker the coating layer is applied. Therefore, the repair coating should always be applied in the biggest possible layer. For repairs of defects it is more suitable to use a repair coating of a darker shade, as a light shade of the coating starts to shine on the gradually darkening substrate (Figure 7.47), unlike a dark shade of the coating, which does not give such a disturbing impression even on light glossy coating (Figure 7.48).

7.9.3 Zinc spray

Zinc spray (Figure 7.49) cannot be recommended for repairs of defects of galvanized coating in any case as it exhibits a very low content of zinc dust and it can only be applied in a relatively thin layer. Local application of the spray results in conspicuous highlighting of the treated place (Figure 7.50). Zinc spray can be used in exceptional circumstances to unify the coating shade on the whole area, from edge to edge, in cases when it is justifiably necessary to repair a cosmetic defect of zinc coating and achieve a decorative effect.

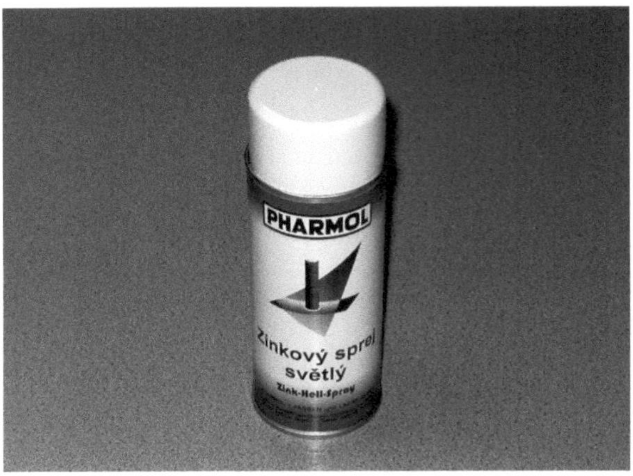

Figure 7.49 Zinc spray.

Figure 7.50 Hot dip galvanized part with zinc spray application after a short exposure in the outdoor environment.

7.9.4 Other defect repair methods

The EN ISO 1461 standard allows application of zinc paste or zinc alloy solder for repairs of zinc coating defects. The suitability of both the materials for a particular purpose must be confirmed by the manufacturer in the technical documentation of the product. When zinc alloy solder is used, the heat accumulated in the galvanized part must be utilized and you should proceed quickly before its temperature drops below the temperature required for solder to melt and at the same time the part should be adequately positioned to make sure the applied solder layer is sufficient.

Within the execution of zinc coating repairs, the manufacturer's technical instructions must mainly be observed.

Service life of hot-dip galvanized coatings

8

The properties of zinc that protect steel from corrosion consist of:

- it creates a sufficiently thick barrier
- cathodic protection in most environments
- ability of zinc to passivate

8.1 Anti-corrosion action of zinc

A galvanized zinc coating is naturally interconnected with the steel substrate as it is produced through a metallurgical reaction between the iron contained in galvanized steel and molten zinc. During this reaction, a barrier consisting of iron-zinc alloy phases grows on the substrate, which reliably separates galvanized steel from corrosive environments. This metallic barrier has a very good resistance to abrasion and it is able to resist erosion for a relatively long time (Figure 8.1). Iron-zinc alloy phases usually manifest a higher hardness than structural steel of standard quality.

Zinc is capable of providing parts made from ferrous metals with very efficient anti-corrosion protection in the long term. The way it protects steel from corrosion results from its physical and chemical properties. In most corrosive environments, zinc exhibits a lower redox potential than iron. If electrically conductive bonding is established between iron and zinc, in the presence of an electrolyte such as moisture, zinc acts as a sacrificial anode. Zinc ions are released from the zinc anode and are attracted to the iron cathode (Figure 8.2). Iron passes into the state of immunity and does not corrode. However, every coating is affected by various integrity defects and pores through which moisture penetrates to the substrate (Figure 8.3). Therefore, the effect of cathodic protection is important even without visible coating disruptions.

Values of the redox potentials of metals differ from each other in various environments and conditions. Under specific conditions it may happen that the polarity of the iron-zinc galvanic cell will change (Figure 8.4). The value of the redox potential of iron drops below that of zinc and in such a solution iron is the first to corrode [47]. Such a situation may occur, e.g., if the electrolyte is water heated to over 65°C. The fact that zinc passes into the state of immunity here and stops protecting iron plays a significant role in using hot-dip galvanized elements for the distribution of hot utility water (Figures 8.5 and 8.6).

An important feature of zinc coating is that on its surface it forms a natural, sufficiently thick and very stable patina that separates the coating from the corrosive

Hot-Dip Galvanizing of Steel Structures. DOI: http://dx.doi.org/10.1016/B978-0-08-100753-2.00008-2

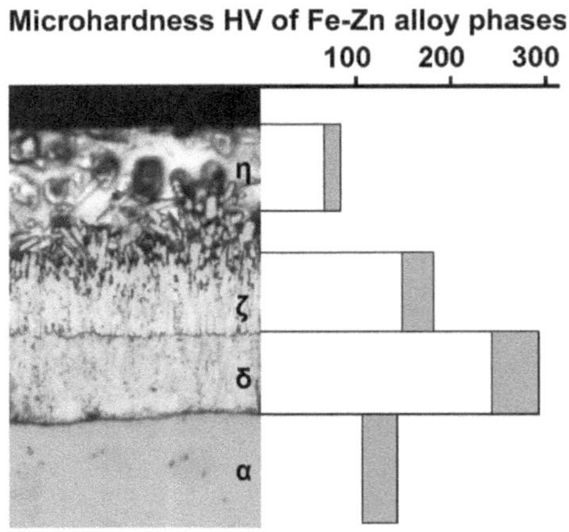

Figure 8.1 Microhardness of Fe-Zn phases.

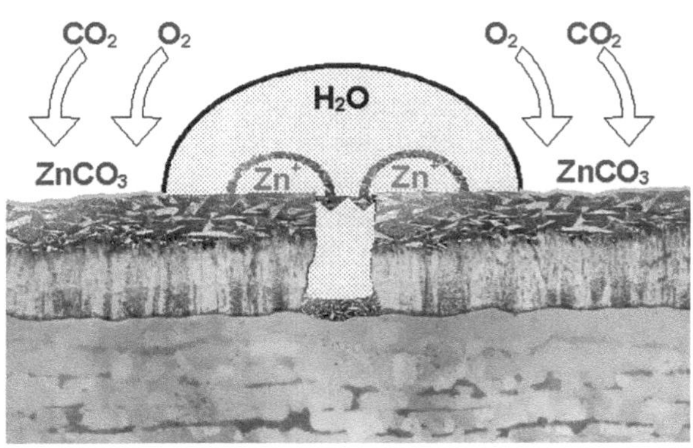

Figure 8.2 Barrier formed by zinc coating, cathodic protection in a pore, and passivation of zinc on the coating surface.

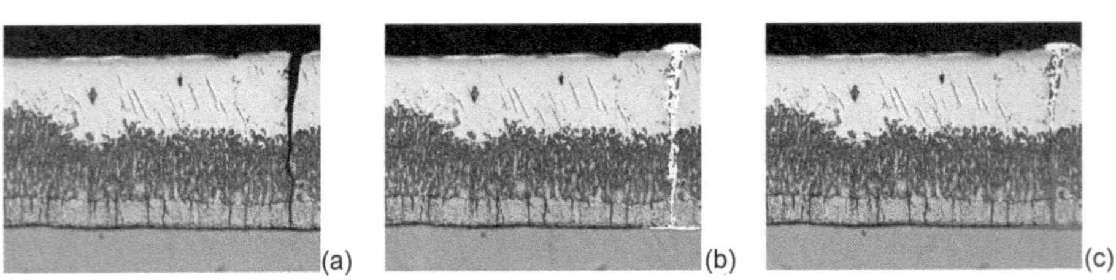

Figure 8.3 Coating affected by defects of integrity (photomontage): a) pore, b) water capillary elevation into a coating pore, c) pores closed by zinc corrosion products.

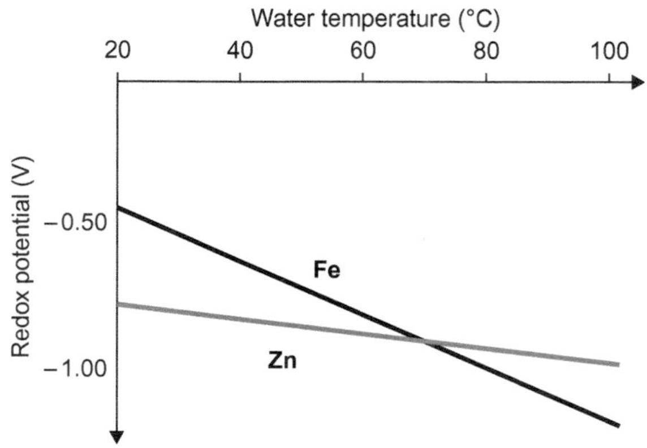

Figure 8.4 Polarity change of the Fe-Zn cell depending on water temperature [47].

Figure 8.5 Point corrosion of the bottom of a galvanized water heater – corrosion tubercles in the places of corroded coating.

environment. The ability to become efficiently passivated is an important characteristic of zinc, which otherwise does not resist corrosion well. A patina, consisting mainly of zinc carbonate, exhibits excellent adhesion and relatively good mechanical and chemical resistance in a neutral to slightly alkaline environment (in an acidic environment it quickly dissolves).

The chemistry of zinc passivation with the production of zinc carbonate (see also Section 7.6) is as follows:

$$ZnO + H_2CO_3 \rightarrow H_2O + ZnCO_3$$
$$Zn(OH)_2 + CO_2 \rightarrow H_2O + ZnCO_3$$
$$Zn(OH)_2 + H_2CO_3 \rightarrow 2H_2O + ZnCO_3$$
$$Zn(HCO_3)_2 \rightarrow H_2O + CO_2 + ZnCO_3$$

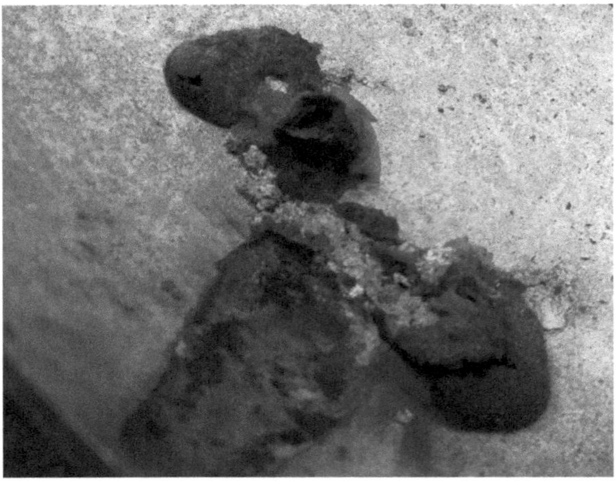

Figure 8.6 Corrosion of a galvanized water heater – crater-shaped deposition of corrosion products of iron indicates an intensive course of the corrosion reaction.

Zinc carbonate is a gray substance insoluble in water. It covers the substrate in a continuous, highly adherent and both mechanically and chemically resistant thin patina layer.

8.2 Atmospheric corrosion

In the course of time, the coating thickness is reduced at a variable rate depending on the corrosivity of the environment. Its useful life is indirectly proportional to the corrosion rate, expressed usually as the amount of corrosion loss in micrometers or grams per unit area. The anti-corrosion protection system of a hot-dip galvanized coating is very efficient if the galvanized part is exposed in a clean atmospheric environment. This means an environment where free access of air to the part's surface is ensured and its possible wetting is temporary and short. The concentration of acidic pollutants in the atmosphere must also be low.

The passivation layer of zinc carbonate is slowly reduced by the action of environmental effects, but it is continuously renewed by zinc contained in the coating. The time for which a zinc coating is able to protect steel from corrosion is derived from the service life of this coating, i.e., the time until it is consumed by corrosion. This time period is directly proportional to the thickness of the applied zinc coating and indirectly proportional to the corrosion rate of zinc in the particular corrosion environment [48]. The values of average annual corrosion losses depending on the corrosivity of the environment are published in a number of technical standards (e.g., EN ISO 12944-2 or ISO 9223). Both of them classify atmospheres in six classes C1 to CX by increasing corrosivity (see Table 8.1).

A sufficiently thick patina layer is created on a zinc coating in the course of a few weeks after its application. Until then the coating has reduced ability to resist

Table 8.1 Corrosion rates of zinc in accordance with EN ISO 9223:2012

Corrosivity class			Corrosion rate (r_{corr})	
			Zinc	**Units**
C1	very low	well-ventilated interiors of commercial, office, and residential premises with a maintained temperature and without the risk of moisture condensation, dry or cool outdoor areas with very low air pollution (deserts, central Antarctica)	$r_{corr} \le 0.7$ $r_{corr} \le 0.1$	[$g.m^{-2}.year^{-1}$] [$\mu m.year^{-1}$]
C2	low	ventilated sports halls, production halls, garages, storage premises without a maintained temperature with clean environment and the possibility of short-term moisture condensation, dry or cold climatic zone with a short wetting period, clean country environment	$0.7 < r_{corr} \le 5$ $0.1 < r_{corr} \le 0,7$	[$g.m^{-2}.year^{-1}$] [$\mu m.year^{-1}$]
C3	medium	environment of industrial halls with medium pollution (food processing plants, laundries, dairies, breweries, mild climate regions with medium and tropic regions with low air pollution, municipal environment, maritime environment with low salinity)	$5 < r_{corr} \le 15$ $0.7 < r_{corr} \le 2,1$	[$g.m^{-2}.year^{-1}$] [$\mu m.year^{-1}$]
C4	high	environment with high frequency of condensation and high pollution, indoor swimming pools, mild climate regions with high and tropic regions with medium air pollution, industrial areas with medium air pollution and maritime atmospheres with medium salinity, aerosol of road salt	$15 < r_{corr} \le 30$ $2.1 < r_{corr} \le 4.2$	[$g.m^{-2}.year^{-1}$] [$\mu m.year^{-1}$]

(Continued)

Table 8.1 **(Continued)**

Corrosivity class			Corrosion rate (r_{corr})	
			Zinc	**Units**
C5	very high	poorly ventilated environment with very high frequency of condensation or high pollution from production processes and mines, mild climate regions with very high and tropic regions with high air pollution, industrial areas with high air pollution and maritime atmospheres with high salinity	$30 < r_{corr} \leq 60$ $4.2 < r_{corr} \leq 8.4$	[g.m^{-2}.year^{-1}] [μm.year^{-1}]
CX	extreme	premises with permanent condensation and high pollution from production processes, regions with very high air pollution, tropical maritime regions with frequent and long wetting and very high salinity, environment of fishing ship with direct exposure to salt water	$60 < r_{corr} \leq 180$ $8.4 < r_{corr} \leq 25$	[g.m^{-2}.year^{-1}] [μm.year^{-1}]

corrosive influences of the environment and being moist. During this time it can be covered by corrosion products – white rust. In normal climatic conditions, effects on the zinc coating by white rust show a temporary character. Climatic influences gradually cause removal of corrosion products from the surface and their gradual replacement by a layer of patina (zinc carbonate). After several months' exposure, the surface that has been affected by white rust acquires the same patina appearance as the remaining coating parts that have not been attacked by corrosion. As with any other product, galvanized coatings are subject to certain restrictive conditions limiting the suitability of their application and way of use. On condition the user respects the guarantee conditions and eliminates possible harmful local influences; galvanized coatings are able to provide steel products with reliable anti-corrosion protection for a very long time. In most applications they represent the most suitable, efficient, and reliable anti-corrosion system.

In many European countries, initiatives from national galvanizers' associations have given to rise to maps of the atmospheric corrosion rate of zinc. These tools are used for quick reference. The zinc corrosion rates in the UK and Republic of Ireland shown in Figure 8.7 [49] were separated into five categories (these categories are not intended to be correlated to the "C categories" of ISO 9223).

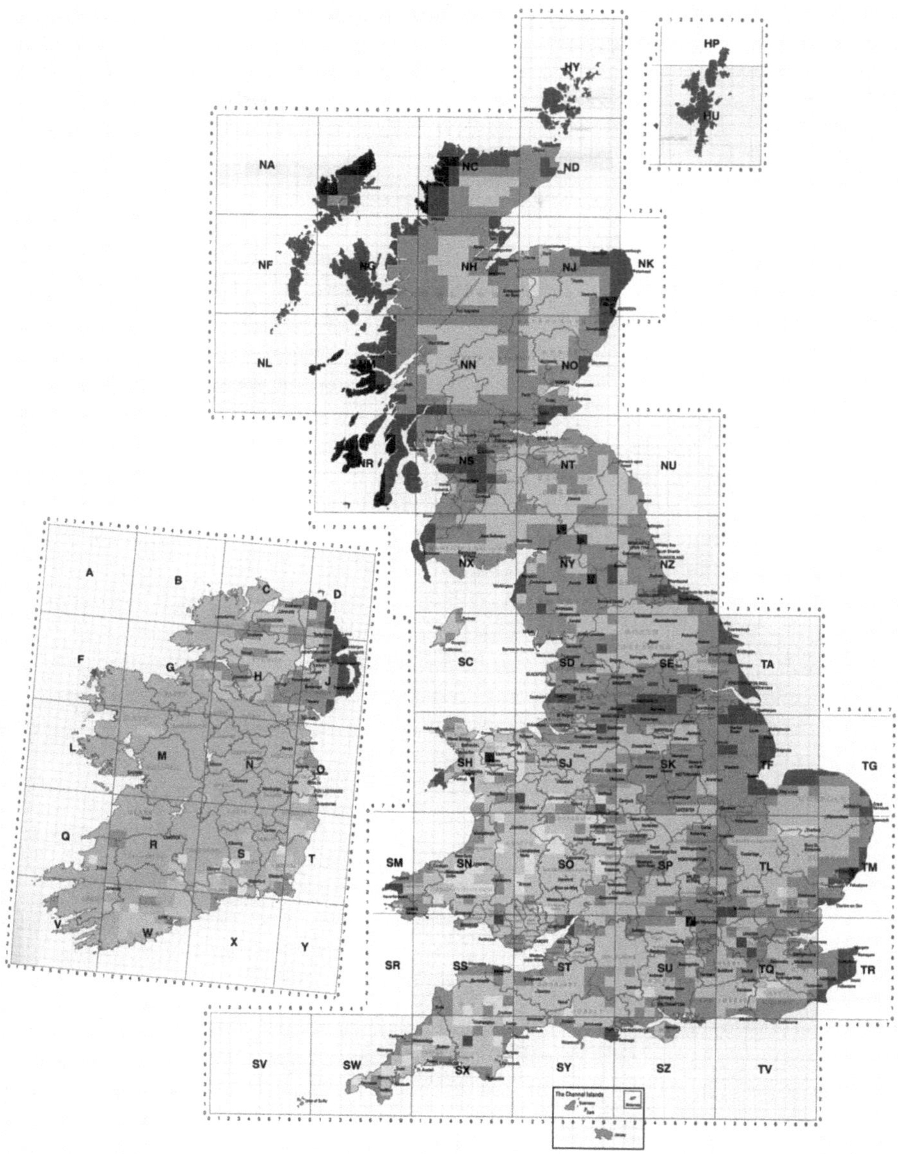

Figure 8.7 The Zinc Millennium Map.
Image is provided courtesy of Galvanizers Association.

Table 8.2 Atmospheric zinc corrosion rates in the UK are represented by five categories

Corrosion category	1	2	3	4	5
Average corrosion rate [μm/year]	0.5	1	1.5	2	2.5
Average life of 85 μm galvanized coating [years]	170	85	57	43	34

In Table 8.2, according to color-coding of each map block, the value of the average annual corrosion losses of zinc due to atmospheric corrosion, i.e., without considering micro-climatic influences, is deduced. The design of a particular anti-corrosion system must also take local influences into account. They should include any major air pollution source in the vicinity, industrial activity, or an unsuitable structural design of the part allowing deposition of solid pollutants in inadequately designed structural elements, etc. (see also Section 9.11).

With knowledge of the corrosion rate of zinc in a particular environment, it is relatively easy to estimate the service life of a galvanized coating. With knowledge of the minimum coating thickness (Tables 7.2 or 7.3) and the corrosion rate of zinc in the particular environment, the expected service life is easy to determine. EN ISO 1461 sets minimum coating thickness values; however, it does not prevent the purchaser agreeing different values with the galvanizing plant. The application of a coating with a required thickness can be influenced by selecting the thickness of the semi-product or by using silicon-killed steel. A thicker coating may also be achieved by blasting the part's surface with a fine sharp-edged abrasive material (see Section 5.4.3).

8.3 Duplex systems

In practice, metallic coatings and coatings of organic painting materials are used to protect steel from corrosion. Coating systems differ from each other in the principle of their anti-corrosion action. Organic painting materials form a barrier after application that separates the substrate from the corrosive environment. The efficiency of such coating systems mainly depends on their permeability and resistance to degradation. In the course of time the organic painting material degrades and becomes more and more permeable to humid air, which penetrates to the substrate through pores in the coating and corrosively attacks it. The coating loss due to erosion is immaterial as corrosion products originating at the boundary between the substrate and the organic coating gradually considerably increase their volume, causing delamination of the coating (Figure 8.8).

In the case of metallic coatings, what must be distinguished is whether the coating metal has a higher or lower electrode potential than iron in the particular environment [47]. An integral metallic coating protects steel regardless of the nobleness of the coating metal as long as the metal is present on the surface. If the metallic

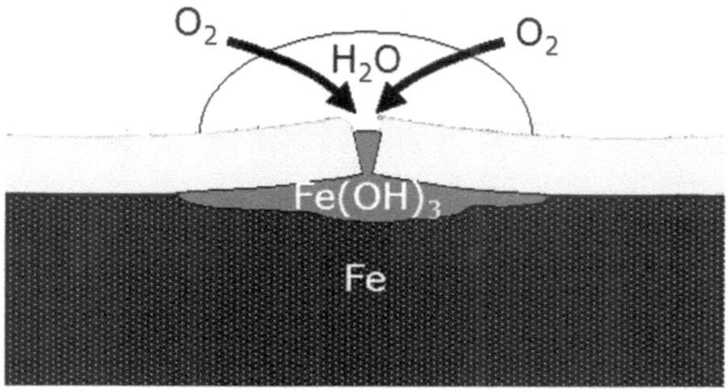

Figure 8.8 Failure in organic coating.

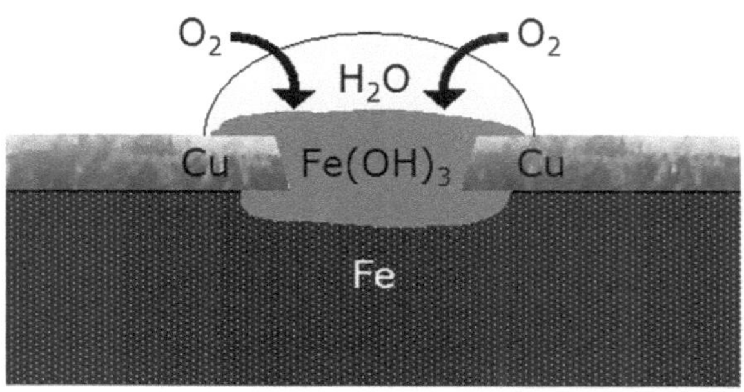

Figure 8.9 Failure in noble metal coating.

coating of a nobler metal is damaged, the coating metal becomes the cathode. In such a place the steel anode preferentially corrodes (Figure 8.9).

A different situation occurs with coating of a less noble metal, as described in Section 8.1. Such coating may be porous or damaged, and in spite of this it provides the substrate with cathodic protection and protects it from corrosion as long as it is present on the surface.

The corrosion rate of the metal depends on the degree of corrosivity of the atmosphere it is exposed to. Zinc carbonate (zinc patina) dissolves well in inorganic acids; therefore zinc is not resistant to acidic environments and very quickly dissolves in them. On the other hand, organic painting materials may be equally resistant to acidic and alkaline environments depending on their composition.

If a hot-dip galvanized coating is covered by organic coating, penetrated moisture causes passivation of zinc, which means that it does not corrode any further. Organic coatings prevent losses of zinc patina at the boundary between the zinc coating and the painting material, and corrosion of zinc close to the patina is stopped (Figure 8.10).

In a duplex system, both the coatings synergistically support each other. The organic coating prevents erosion of the patina covering zinc and protects it from corrosion. At the same time, passivated zinc does not reduce adhesion the organic

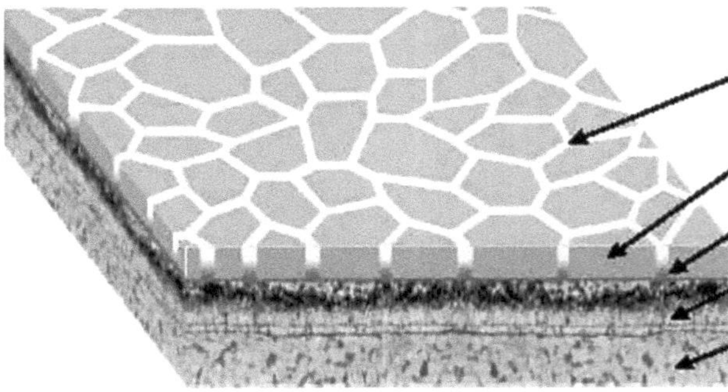

Figure 8.10 Duplex system.

Figure 8.11 Organic coating on steel.

coating layer to the substrate as it would if the organic coating were directly applied onto steel (Figure 8.11). The organic coating is protected by zinc from the bottom, which extends its service life by several times. The system is resistant to a corrosive environment until the organic painting material degrades to such an extent that it starts to disintegrate. It is only then that corrosion of the zinc coating occurs. Therefore, the useful life of a duplex coating system is significantly longer than the sum of the useful lives of both the coatings if they were applied separately. The literature indicates that the durability of a duplex system is 1.5 to 2.3 times higher than this sum [15]. The physical life of structures treated this way is often equal to their technical-economical life.

Duplex systems are generally applied on the substrate in sequence by two different coating contractors. Therefore, the hot-dip galvanized coating must be properly prepared for application of the organic painting material or powder coating. This operation is carried out by the organic coating contractor, who accepts responsibility for the final result on accepting the order. The galvanizer may be asked not to cool the galvanized part in water after the application of the zinc coating or not to treat the coating against white rust formation (see also Section 7.6). The galvanizer must then recommend the purchaser of the surface finish a suitable defect repair method. Defects in the zinc coating should not be repaired by repair paint with a high content of zinc dust (see Section 7.9.2), which may penetrate through the applied organic coating. A suitable method of surface pre-treatment for application of the organic painting material is rough removal of uneven places and lumps on the zinc coating with a file with a rough cut and subsequent sweep blasting with very fine sharp-edged abrasive agent or re-grinding with sand paper. The purpose of the sweep blasting is not only to achieve the required degree of surface cleanness but also the necessary profile for anchoring of the organic painting material or powder coating on the smooth zinc coating. Galvanized coatings applied on low reactivity steels with a low content of silicon (up to about 0.03%) exhibit a fine structure and have a very good mechanical resistance. They are more suitable for sweep blasting than thick coatings applied on silicon-killed steel. Reduced adhesion to the substrate is a characteristic of very thick coatings. They easily delaminate even if the sweep blasting is done very carefully. After application of the galvanized coating an inter-operation inspection must be performed to prove conformity of the galvanized coating to the requirements of EN ISO 1461. A part of the inter-operation (input) inspection carried out by the organic coating contractor should also be evaluation of suitability of the galvanized coating for sweep blasting. In this context, the purchaser of the surface finish must select suitable steel grades for parts designed to be applied with a duplex system.

Proper design principles for hot-dip galvanizing

9

Hot-dip galvanized coatings are applied in such a way that after the necessary chemical treatment of the surface, the articles are immersed in the zinc melt at a temperature of approximately 450°C, which ensures that the zinc coats the external as well as the internal surfaces. Commercial galvanizing plants equipped for batch galvanizing using of the modern "dry process" technology use the jig method of processing. The parts to be galvanized are suspended on a special jig in a suitable position, in which they remain while they pass through the entire production process (Figure 9.1). The structural design, fabrication, and workmanship of the articles to be galvanized must be adapted to this process [50].

When designing parts to be hot-dip galvanized, consideration must be given to the aspects that are necessary for the production of a high quality zinc coating on the whole (i.e., external and internal) surface. The designer must respect special principles without which high quality galvanizing would not be possible.

9.1 Drain holes

Molten zinc must be given access to the entire surface of the article as it remains stationary on the jig. Cavities and pockets on the article to be galvanized must have drainage (Figure 9.3) and vent holes (Figure 9.4) of sufficient size to allow the part to be quickly immersed in the zinc bath. At the same time any pockets and corners should not retain pre-treatment solutions, zinc, or air bubbles, which may cause galvanizing defects (Figure 9.8).

The designer must ensure that not only cavities but also all open pockets and corners are provided with suitable drainage and vent holes. During immersion of the article, the molten zinc must be able to completely wet the entire surface and none of it should be retained during removal (Table 9.1). Vent holes may be one order of magnitude smaller size than drain holes. Large enclosed sections require an especially large opening for ingress of molten zinc (Figure 9.2).

The part must be designed in such a way that it can be immersed in the zinc melt as quickly as possible. As the specific weight of steel is slightly higher than that of zinc, in the bottom part of the article at the inlet side, sufficiently sized drainage holes must be incorporated so that slow filling of the cavities with molten zinc should not prevent the required quick immersion of the part.

Unsuitably located and improperly designed drainage holes will not correctly fulfill their function (Figure 9.5). A hole drilled perpendicular to the main plane of the frame causes the formation of pockets that retain chemical pre-treatment solutions in

Hot-Dip Galvanizing of Steel Structures. DOI: http://dx.doi.org/10.1016/B978-0-08-100753-2.00009-4

Figure 9.1 Galvanizing jig.

Figure 9.2 For successful galvanizing the cistern must be fitted with properly sized holders (lugs) for suspension and a hole for quick filling of the cavity with liquid metal and its subsequent emptying and cleaning of ashes.

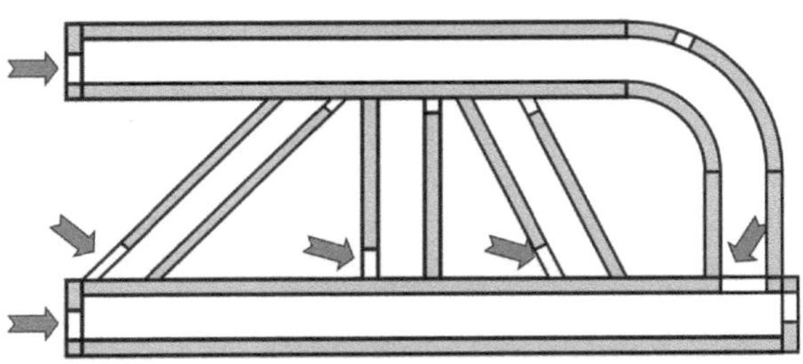

Figure 9.3 Properly located and sized drain holes for the inlet and outlet of pre-treatment baths and liquid zinc.

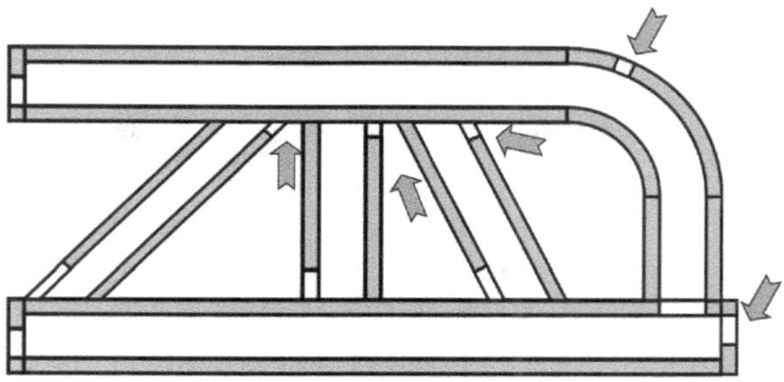

Figure 9.4 Cross venting at the opposite side of the drains.

Table 9.1 **Dimensions of drain holes necessary per 1 meter of profile length [50]**

Profile cross-section			Minimum Ø of a drain hole for the number of holes		
◎	▣	▭	1	2	3
15	15	20 × 10	**8**		
20	20	30 × 15	**10**		
30	30	40 × 20	**12**	**10**	
40	40	50 × 30	**14**	**12**	
50	50	60 × 40	**16**	**12**	**10**
60	60	80 × 40	**20**	**12**	**10**
80	80	100 × 60	**20**	**16**	**12**
100	100	120 × 80	**25**	**20**	**12**
120	120	160 × 80	**30**	**25**	**20**
160	160	200 × 120	**40**	**25**	**20**
200	200	260 × 140	**50**	**30**	**25**

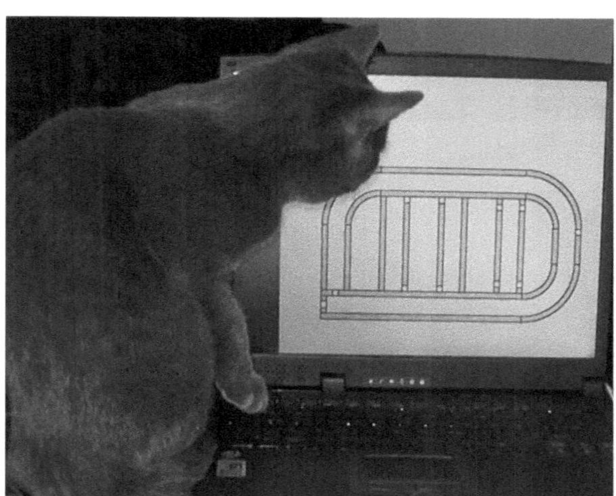

Figure 9.5 Such a design is only "good for a cat."

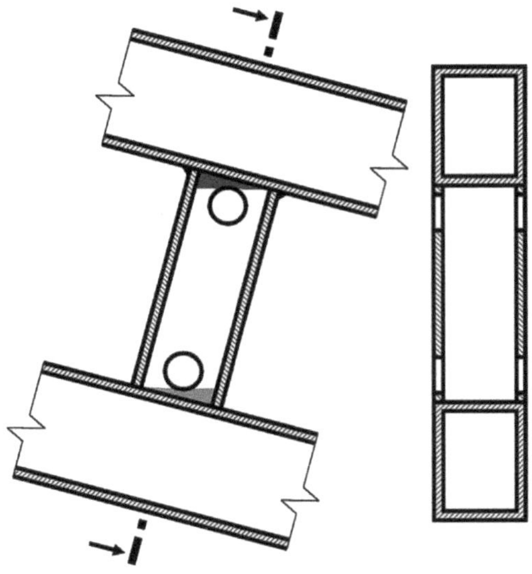

Figure 9.6 Insufficient drain holes.

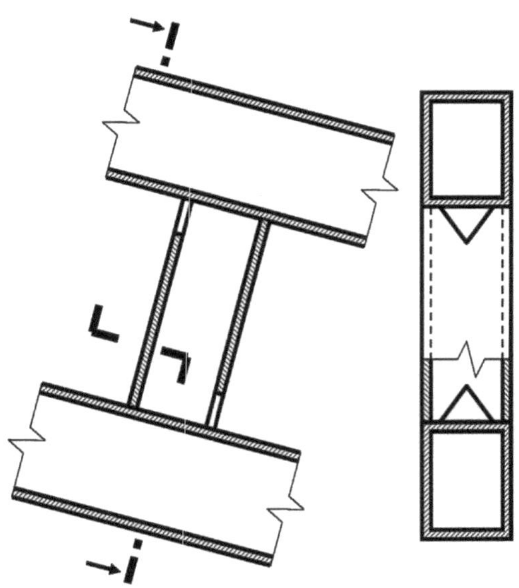

Figure 9.7 Properly designed drain holes.

the bottom part and subsequently also zinc. An air bubble is trapped in the top part, which will create a non-wetted surface that will not be coated (Figure 9.6). The axes of drain (and vent) holes must lie on the vertical main plane of the frame. In this example, the proper provision of drainage and venting involves cutting or grinding the profile end in such a way to avoid creation of a valley (Figure 9.7).

Figure 9.8 Unsuitably designed venting prevents galvanization of the whole surface of the part.

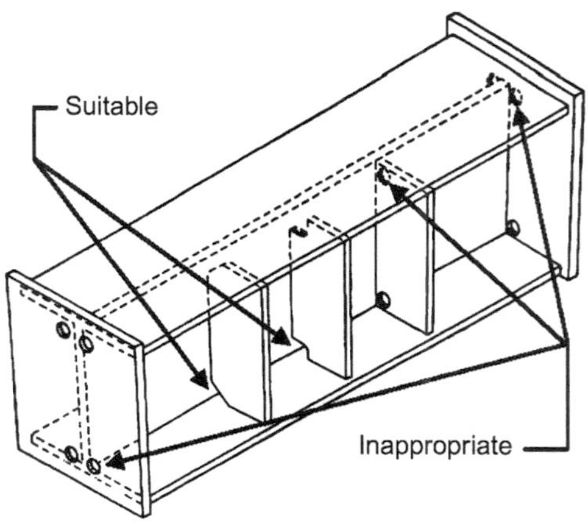

Figure 9.9 Drain holes recommended in accordance with EN ISO 14713-2.

Figure 9.9 shows several examples of drain hole design as recommended by the EN ISO 14713-2 standard. However, the drilled holes that are indicated in the rib in the right part and in both the flanges are not a good solution as they give the possibility of formation of a valley, which may lead to galvanizing defects (Figure 9.8). Examples of a possible design of drain holes of feet (heads) of columns made of rolled profiles are shown in Figure 9.10.

9.2 Variable thermal stress during hot-dip galvanizing

The temperature of the zinc melt is approximately 450°C. The part of the galvanized article that is immersed is heated up to this temperature and expands while

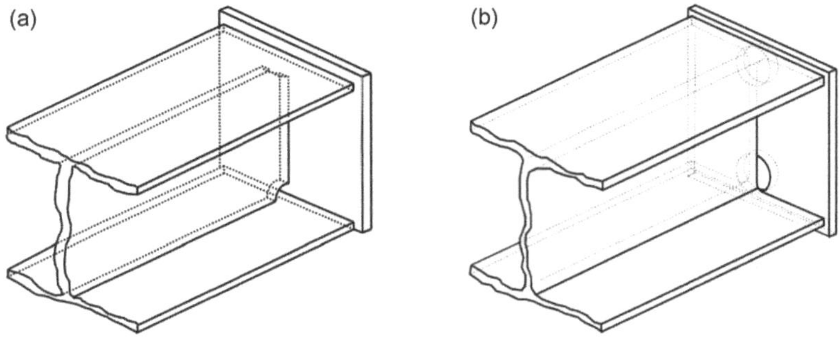

Figure 9.10 Examples of possible drain hole designs: a) by milling (burning) in a profile post, b) by drilling (burning) in a flange.

Figure 9.11 Releasing the residual stress introduced into the part by an incorrect welding process.

the part above the surface is relatively cold and only starts to expand after a delay. In rigid structures this may lead to shape changes and sometimes even to damage to articles or fabrications (see Section 10.6). Galvanized articles must not contain hollow, closed elements that can explode on heating to the galvanizing temperature (see Section 9.9).

The zinc melt temperature of 450°C is close to the temperature for reduction of residual stress (Figure 9.11) or to tempering temperatures.

Hollow bracing elements with insufficient drain holes may hinder the quick immersion of trusses, which may consequently lead to their damage (Figure 9.12). The rigid structure of the truss does not allow free expansion of the elements, which occurs due to uneven heating.

Figure 9.12 Trusses with hollow bracing elements.

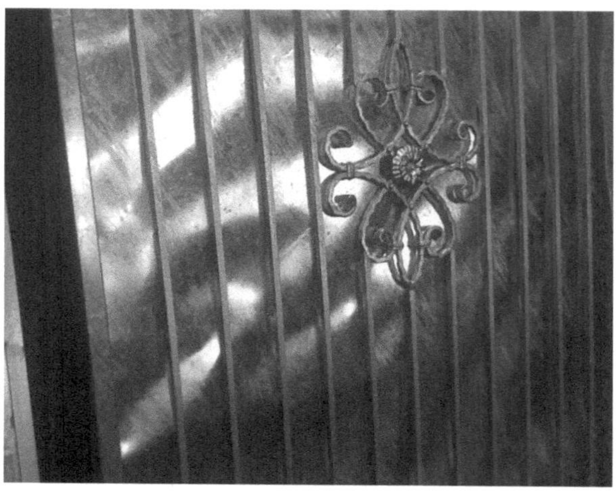

Figure 9.13 Thin metal sheets always corrugate during hot-dip galvanizing.

Another example of an unsuitable structure is a combination of materials with variable section thicknesses (e.g., thin panel sheets within a gate made of hollow profiles). Such structures are generally subject to material deformation (Figure 9.13). A good result can be achieved through creasing or cut-outs in the metal sheets and by separate galvanizing of frames and panel sheets (Figure 9.14).

9.3 The galvanizing plant must be allowed to suspend parts

A contact spot will always remain where the article contacted the jig and this will exhibit a discontinuity in the zinc coating. Burrs surrounding the contact spot must be smoothed with a file and anywhere that the coating has peeled off must be

Figure 9.14 Structural treatment against sheet deformation.

Figure 9.15 Imprint of a suspension fixture (chain) on the galvanized coating.

repaired (e.g., using a special repair paint with a high content of zinc dust in the dry film). It is a good idea to fit a part intended to be hot-dip galvanized with lifting lugs. If there are no suitable lugs there is a risk that the galvanized surface will be affected by an imprint of the suspension fixture (Figure 9.15).

9.4 Clean metallic surface

Parts that are sent for hot-dip galvanizing must not contain any coating, marking, or impurities that are not soluble in water (see also Section 7.7.3) and that cannot be removed by pickling in hydrochloric acid (Figures 9.16 and 9.17).

A necessary precondition for the formation of a high-quality iron-zinc alloy coating is a clean metallic surface. However, it is not recommended to sand-blast parts

Figure 9.16 Missing coating in a place where a sticker was left.

Figure 9.17 Unacceptable color marking of parts to be hot-dip galvanized.

designed to be hot-dip galvanized, unless this is done intentionally to increase the thickness of the applied coating, to clean the surface of a thermal cut, to remove welding slag, or when cast iron is hot-dip galvanized. On articles to be hot-dip galvanized (and in their cavities) there must not be any contaminants that cannot be removed using a normal chemical pre-treatment process.

Contaminants in cavities that are insoluble in the pickling solution represent an increasing risk of defects as they act as crystallization cores on which hard zinc crystals preferentially grow. In a mixture with zinc melt they form mushy matter that causes the occurrence of unsightly effluences around drain holes (Figure 9.18). A similar effect is produced by pickling solution retained in a pocket or in a corner of a part with improperly designed drain holes.

Figure 9.18 Seepage of dross from a cavity with an insufficiently sized drain hole.

Figure 9.19 Porous weld.

9.5 Welds

Good-quality coating can only be applied on a weld produced in a workmanlike manner. Release of internal stress introduced to the article by an incorrect welding process also contributes to deformation (Figure 9.11). Welds with incomplete penetration (Figure 9.19), using an unsuitable separation agent (see Section 7.7.3), or adhering slag residues after welding with packed electrodes (knocking off is an insufficient process – see Figure 9.20) lead to coating defects.

Figure 9.20 Uncleaned slag.

Figure 9.21 Bulking of coating in a hole and a burr in a place of suspension.

9.6 Holes for pins and screws

If no subsequent calibration of openings for screws or pins is expected after galvanizing, a sufficient tolerance must be considered (Figure 9.21), which should be about 2 mm.

Threads on batch galvanized parts should be protected from coating application in a suitable manner. Inner threads of larger diameters can be coated with a suitable lacquer. Small diameters can be protected with screws (Figure 9.22) whose shank is flush with the surface on the opposite side (they must not protrude from the threaded hole), then they can be unscrewed easily without application of excessive force. An outer thread can be coated with a suitable lacquer or protected with a special tape (Figure 9.23). After galvanizing of the article, the thread should be cleaned with a wire brush.

Figure 9.22 Inner thread protection.

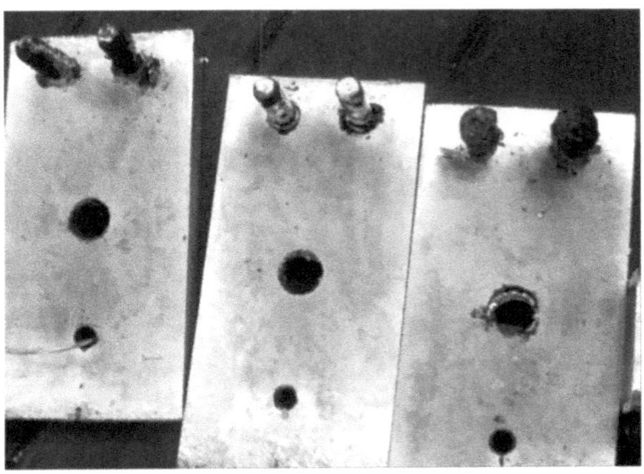

Figure 9.23 Outer thread protection.

9.7 Three-dimensional components

Parts designed to be hot-dip galvanized should, if possible, be designed to be two-dimensional. Three-dimensional parts (Figure 9.24) are problematic in terms of achieving a suitable inclination of all structural elements to ensure outflow of excessive zinc during removal of the part from the zinc bath. Horizontal elements are often affected by bulking of the coating and sharp spikes at the trailing edge (Figure 9.25). Furthermore, if a voluminous light part is dipped, utilization of the galvanizing bath is uneconomical and the galvanizing costs increase disproportionately.

Figure 9.24 A light voluminous part means uneconomical galvanizing.

Figure 9.25 Bulking of the coating at the trailing edge of a three-dimensional element where a sufficient inclination for the outflow of liquid zinc during removal from the zinc bath could not be achieved.

9.8 Overlapping joints

An acidic pickling solution or flux may easily penetrate into an unsealed gap of an overlapping joint during chemical pre-treatment. Due to its high surface tension, zinc does not wet the steel surface, does not penetrate into a gap, and usually does not even fill the gap. Later, when subjected to moisture, aggressive chlorides react with the steel inside the gap and then emerge onto the surface, which gets stained by "foreign rust" in the vicinity of the gap (Figure 9.26). The galvanizer should be consulted before sealing of any overlapping joints with continuous welds and an

Figure 9.26 Pollution on the coating near a gap caused by "foreign rust" seeping from the gap.

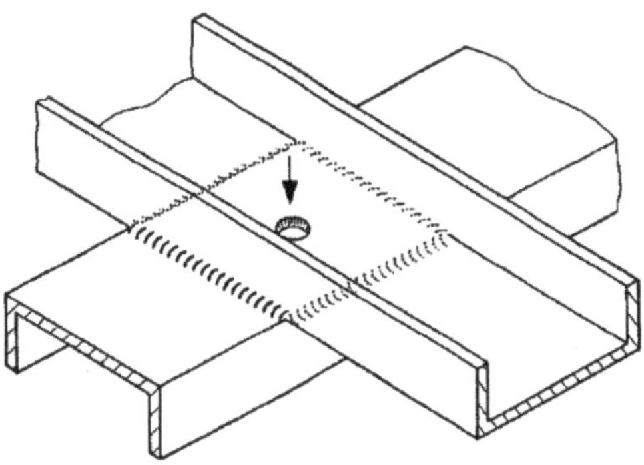

Figure 9.27 Design of overlapping joints as recommended by EN ISO 14713-2.

assessment of their safety should be made. In case of a larger joint area there is a risk of buckling or tearing and therefore it is recommended that one of the elements be fitted with a vent hole (Figure 9.27) or to leave out a part of the weld as a gap vent. The vent must then be protected from penetration of the pickling solution by closing with a suitable sealant. The weld must be waterproof so that the pickling solution should not get into the gap through leaky places; otherwise there is a danger of seepage of chlorides during the immersion of the part into the zinc melt. The stained surface then remains uncoated (Figure 9.28).

9.9 Closed cavities

All cavities on a galvanized article must be fitted with drain and vent holes. The pressure of gas enclosed in the cavity and heated to the zinc melt temperature

Figure 9.28 It is difficult to produce a waterproof weld around the whole perimeter of a too large lap joint surface. Chlorides that penetrated into the gap during the pickling boil out during the galvanizing, stain the vicinity of the vent hole, and cause a zinc coating defect.

Figure 9.29 Element of a hollow closed profile after an explosion in the zinc bath.

will cause an explosion (Figure 9.29). This is dangerous not only for the article itself and the equipment of the galvanizing plant but most importantly for the operators of the galvanizing workplace.

Buckling caused by expanding gas may even occur in the case of correctly dimensioned venting if the pickling solution is retained in the cavity, as this evaporates and expands at the temperature of 450°C (Figure 9.30).

Figure 9.30 Buckling of a box-like part wall caused by retained pickling solution.

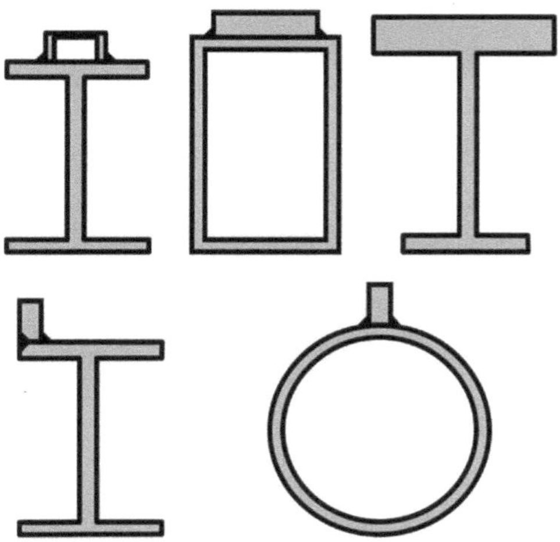

Figure 9.31 Asymmetric parts.

9.10 Asymmetric parts

Asymmetric parts (Figure 9.31) are heated up unevenly during gradual immersion. The stress in galvanized parts reaches the yield point of the material and therefore asymmetric parts will be deformed to an increased extent (Figure 9.32).

9.11 Deposition of dirt and debris

The structural design of the steel structures must avoid deposition of dirt and other debris because this represents a significant local corrosion initiator during long-term exposure and reduces the durability of the protection (Figure 9.33).

Figure 9.32 Deflection of asymmetric structural elements.

Figure 9.33 Increased corrosion load due to deposition of dirt in a corner.

9.12 Moving assemblies

In the case of elements that should be movable with respect to each other, it is essential that the part does not get connected with its counterpart. Excessively tight seating of a pin in the hole will cause the mating surfaces to remain uncoated. On the other hand, too loose a seating leads to soldering (Figures 9.34 and 9.35). Therefore, assemblies must be sent for galvanizing in the disassembled state. Individual parts are galvanized separately and only assembled after being coated. If necessary, mating holes can be calibrated before the assembly. As regards pins, the galvanizing plant must be notified of the functional surfaces, which are not to be coated. A suitable suspension position for galvanizing can reduce the risk of the coating accumulating near the trailing edge.

Figure 9.34 Pulley with a pressed-in bearing.

Figure 9.35 An antifriction bearing loses its function after hot-dip galvanizing.

Maintaining the integrity of hot-dip galvanized steel structures

10

Hot dip galvanizing is done by immersing the parts to be galvanized in a zinc bath with a temperature higher than the melting point of zinc (419.4°C). Batch galvanizing is generally carried out at temperatures of 445 to 455°C. The articles immersed in the zinc melt gradually, so the section that is in contact with the zinc melt is heated up relatively quickly while the temperature of the section above the surface is at around the ambient temperature. This temperature inhomogeneity is also aggravated by gradual heating of the material from the surface toward its core. A part that always contains residual stress from production is thus exposed to other variable stress, which can locally reach the yield point of the material. Steel parts get deformed during hot-dip galvanizing [51]. In a system represented by an unevenly heated article, the deformation energy is produced by thermal expansion of the material; when the bath temperature has been reached, expansion of the steel ceases.

In addition to variable stress, an important role is also played by the fact that the material characteristics of steel exposed to an elevated temperature are changed compared to their nominal values. The values of the elastic modulus, ductility, toughness, yield point, and ultimate strength are all temporarily reduced.

The most significant negative factors increasing susceptibility to the occurrence of cracks during hot-dip galvanizing include:

- harmful influence of hydrogen
- tempering brittleness
- cold working and strain age embrittlement
- precipitation hardening (steel aging)
- weld and the heat-affected zone
- structures with an inappropriate design or poor quality
- temperature gradient from surface to core of the article
- bimetallic thermal stress
- rate of accumulated elastic energy
- stress history of alternating tensile and compressive stress

In this chapter, the term "bimetallic effect" or "bimetallic stress" is used to describe the effects of differential thermal expansion and consequent stresses created by two parts of a fabrication that have different coefficients of thermal expansion as a result of features such as section thickness.

Hot-Dip Galvanizing of Steel Structures. DOI: http://dx.doi.org/10.1016/B978-0-08-100753-2.00010-0

10.1 Defects caused by hydrogen

Galvanized parts must have a clean metallic surface to allow the metallurgical reaction during the application of the coating to take place. Therefore they are usually pickled in hydrochloric acid. During the reaction of iron with hydrogen chloride, nascent hydrogen is released, which moves to places where its partial pressure is lower, in the particular case of steel (Figure 10.1, see also Section 3.6.3).

10.1.1 Hydrogen embrittlement

Atomic hydrogen travels interstitially in the crystal lattice of iron and after the completion of pickling it gradually escapes from the steel to the atmosphere or gets stuck in places where it can recombine (in "hydrogen traps"). Recombined hydrogen is not capable of any further diffusion and in the places where it was captured it causes a gas pressure amounting to several hundreds of MPa. Thus it can impair the material characteristics of steel [29]. It therefore follows that parts that have a higher frequency of hydrogen traps represent a higher risk. This is the case for elements of parts that have been materially reformed by shaping as well as materials affected by the presence of dislocations, discontinuities, inclusions, folds, and especially welds and their heat affected zones.

10.1.2 Hydrogen corrosion

Hydrogen may bond with free carbon in steel, producing methane, which is not capable of any further diffusion. It accumulates in places of its occurrence, usually at grain boundaries where carbon previously precipitated, e.g., from residual austenite [51] in welds.

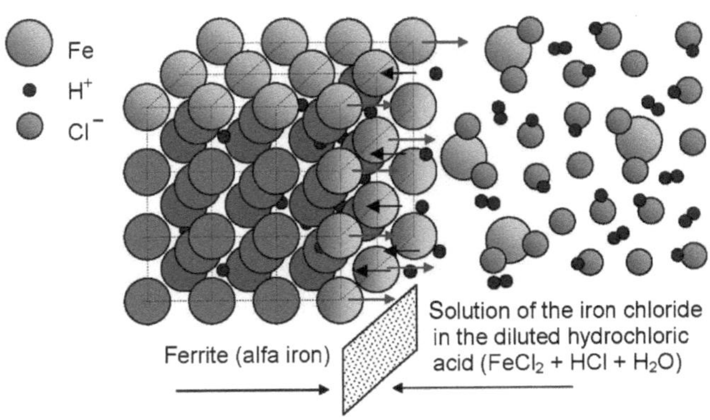

Figure 10.1 Principle of steel hydrogenation during pickling.

10.2 Tempering brittleness

During hot-dip galvanizing tempering brittleness is caused by the martensitic transformation during cooling of the unstable austenite, which is characterized by a higher solubility of carbon than ferrite. At grain boundaries carbon precipitation can occur [51]. It is true that structural steel has a ferritic structure in its natural conditions; however, the presence of austenite cannot be excluded in welds and in heat-affected zones. When hot-dip galvanized parts with residual austenite (welds) present are pickled in acid before application of the coating, tempering brittleness may be accompanied by hydrogen corrosion (see Section 10.1.2) and the parts could disintegrate spontaneously (Figure 10.2).

10.3 Cold working and strain age embrittlement

Strain age embrittlement of steel and especially disruption of its crystalline structure occur at cold worked articles. Material cohesion at grain boundaries is impaired, which is manifested by worsening of material characteristics in the bending place [29]. The material becomes brittle and, in extreme cases, mere contact with molten metal without the action of external forces may cause spontaneous failure of the part (Figures 10.3 and 10.4). With small bending radii the yield strength of steel may be exceeded while a defect created inside the material represents a hydrogen trap, and recombined hydrogen (see Section 10.1) extremely increases its pressure and may cause disruption of integrity of the part. Cold forming also contributes to acceleration of the steel aging process (see Section 10.4).

Figure 10.2 Galvanized part affected by tempering brittleness.

Figure 10.3 Cold forming – view of a disrupted part.

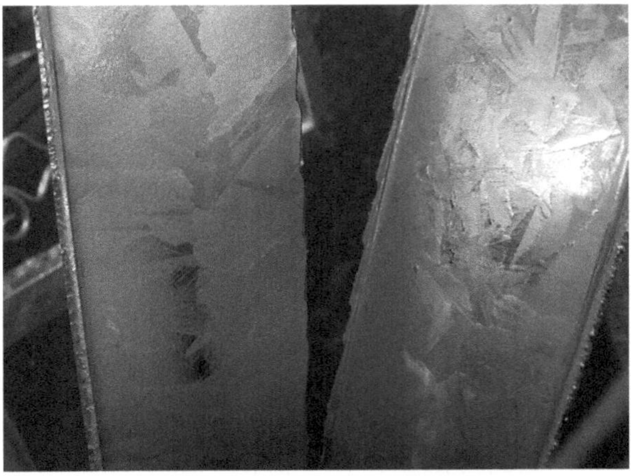

Figure 10.4 Detail of integrity disruption of a galvanized part due to precipitation hardening.

10.4 Precipitation hardening

Precipitation hardening or steel aging is a phenomenon that is manifested by an increase of the transitional temperature to brittle fracture [29]. This is caused by decomposition of the oversaturated solution of nitrogen in steel, which gets into the steel from the furnace atmosphere. An oversaturated solid solution of nitrogen is generated by rapid cooling from the temperature at which this solution is stable. Later, under suitable conditions, diffusion causes precipitation of nitrogen into

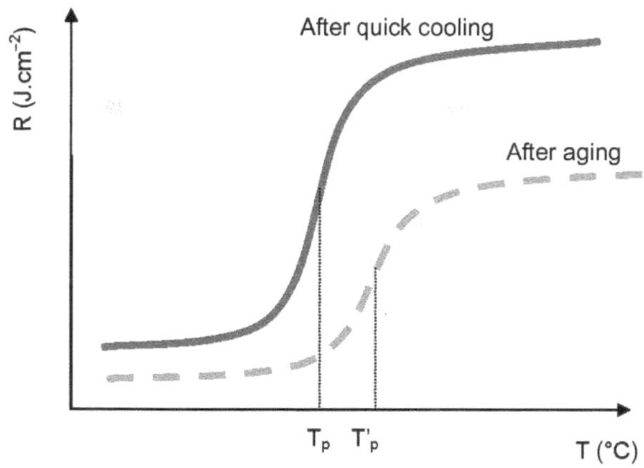

Figure 10.5 Change of the transition temperature to brittle fracture due to steel aging [29].

places of crystal lattice disruption. There, nitrogen disturbs material cohesion and the notch toughness of steel is reduced, the steel becoming more sensitive to tension concentrators. Its transition temperature to brittle fracture T_p rises (Figure 10.5). Precipitation disintegration of oversaturated ferrite may be substantially accelerated by cold forming, which leads to an increase of the number of dislocations.

Steel aging leading to its embrittlement is caused by free nitrogen. The share of free nitrogen in steel can be reduced by its alloying with elements with a high affinity to nitrogen. Most commonly this is aluminum. As the recommended content of aluminum for binding free nitrogen, the limit proportion of atomic masses of $Al/N \geq 2$ is usually mentioned. Other elements able to bind nitrogen, e.g., vanadium (V), titanium (Ti), and niobium (Nb), also meet this requirement. Modern steels show relatively good resistance to aging.

10.5 Welds

A step change of material characteristics occurs in a weld. This negative influence is additionally multiplied by various defects of welds or disturbances occurring in the heat-affected zone. Every weld, regardless of the quality of its execution, is a stress concentrator. Rigid structures, especially trusses of box-shaped frames, are at higher risk of occurrence of cracks (Figure 10.6). The designer's plan of the sequence of variable thermal stress during galvanizing is important for the design of the part. Stresses caused by uneven heating during galvanizing usually have a completely different character from the operational stresses to be experienced by the parts (Figure 10.7).

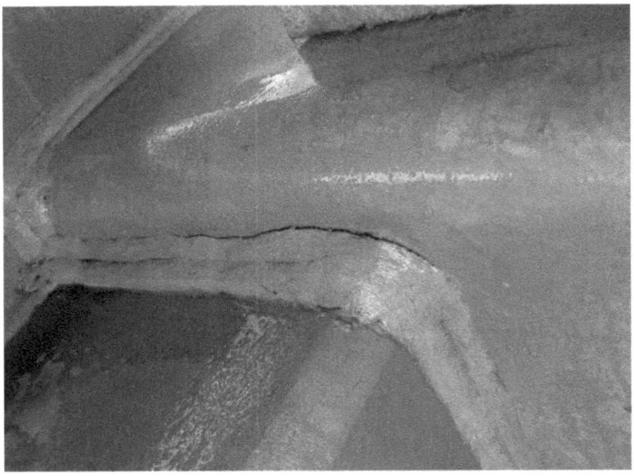

Figure 10.6 Crack in a weld.

Figure 10.7 Poor quality workmanship of a part.

10.6 Structure with an unsuitable design or poor quality

Parts intended to be hot-dip galvanized must be designed in such a way that will enable their free expansion during gradual, uneven heating in the course of immersion in the zinc melt. Slowly filling hollow elements with unsuitable drain holes may cause problems. Stress concentrates in notches. Negligent workmanship of welds on a part designed to be hot-dip galvanized represents a considerable increase of the risk of disruption of its integrity (Figure 10.8). When preparing the structural design and sizing of a part, the designer must not only be limited to its final function and operational load but must also take into account the variable stress that the part is exposed to during galvanizing.

Figure 10.8 Underestimated variable stress during hot-dip galvanizing.

10.7 Thermal stress caused by thermodynamics of hot-dip galvanizing

Stress caused by thermal effects during gradual immersion of the part in the liquid metal reaches the usual yield point of structural steels [51]. An increased risk of occurrence of cracks during hot-dip galvanizing is present in the case of rigid structures made of steels of higher grades, which can be loaded to a higher variable thermal stress due to their higher yield point [51].

10.8 Liquid Metal Assisted Cracking (LMAC)

The hypothesis referred to as Liquid Metal Assisted Cracking (LMAC) is based on the premise that atoms of some liquid metals with a low melting point and relatively high affinity to steel are able to penetrate into it along grain boundaries where they disturb cohesion. Decelerated brittle fracture is characteristic of the LMAC phenomenon (Figure 10.9). Observations [42] have shown that there is an increased LMAC risk in the temperature range between T_e (sudden toughness drop in temperature) and T_r (recovery temperature) (Figure 10.10). What is also interesting is the temperature-dependent decrease of the yield point of structural steel (Figure 10.11). Contribution of tensile stress or other unfavorable influences may then lead to the occurrence of material defects.

The LMAC issue is related to hot-dip galvanizing of large parts designed mainly for building structures. A great boom in using hot-dip galvanizing for steel structures only happened when the formerly insufficient dimensions of galvanizing baths ceased to be a limiting condition for more widespread use. High, slim beams terminated with end plates up to the half of the web plate height

Figure 10.9 LMA crack.

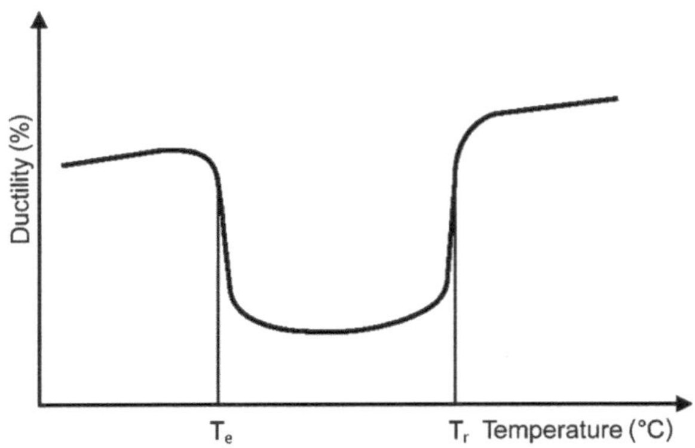

Figure 10.10 The curve shows the course of dependence of steel ductility on its temperature [42].

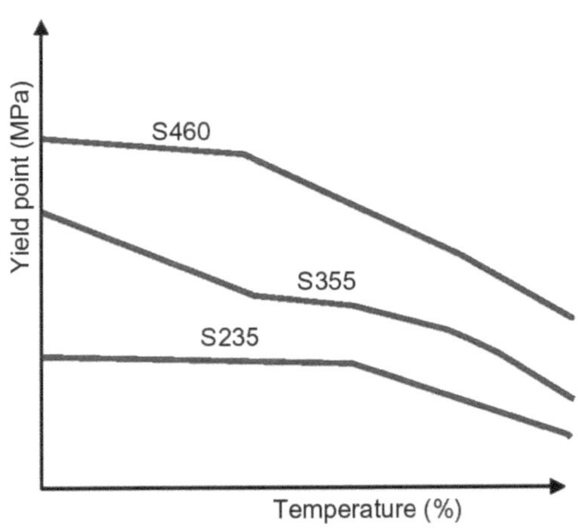

Figure 10.11 Temperature dependence of the yield point of elected structural steel grades [42].

Figure 10.12 Crack coming from a notch in the place of flange termination in the half of the web plate height [23].

Figure 10.13 Rigid, tough structure – general view.

(Figure 10.12) or rigid truss structures (Figures 10.13 and 10.14) can be considered as typical examples of structures prone to cracking in this way.

10.8.1 Bimetallic effect on a structural beam

Variable temperature stress caused by the bimetallic effect during gradual immersion of the part into liquid metal reaches the yield point of the material. Parts made of steel of a typical grade (with a low yield point) are less stressed (relax sooner) during hot-dip galvanizing than parts of high-grade fined-grained steel with a high yield point (able to carry higher stress). The stress level in a part during galvanizing is proportional to the yield point of the steel used.

The increased risk of occurrence of cracks during galvanizing of parts made of higher yield point steel can be explained by an analysis of the thermal influences on a body. During gradual immersion a bimetallic action occurs because there are two mutually coupled elements of the same material but with unequal temperatures

Figure 10.14 Rigid, tough structure – detail with LMA cracks.

[25, 51]. Articles are subject to considerable thermal stress during galvanizing. After reaching the material yield point, the deformation of the galvanized part remains within the plastic range and is deformed to a certain extent (usually acceptable). Structural steel of typical quality (e.g., S235 grade) is characterized by a relatively low yield point and plastic deformation occurs at relatively low stress. The elevated temperature that the steel is exposed to during hot-dip galvanizing causes accelerated creep and the material relaxes. On the other hand, a part made of high-quality fine-grained structural steel (e.g., of the S490 grade), which is characterized by a relatively high yield point, is subject to higher forces. Such structural parts face higher stress caused by heat expansion. Therefore, they are exposed to a higher risk of crack occurrence. Damage occurs at the moment of critical accumulation of energy in a place with a high stress concentration (on the microscopic level) – in places of frequent imperfections in welds and in heat-affected zones of welds (precipitation of nitrides, phosphides, silicates, and carbides at grain boundaries, micro-cracks caused by spatial stressing during shrinking of the weld metal, weld execution defects, etc.), however, generally before achievement of the macroscopic yield point of the material.

Figure 10.15 indicates a simple situation when the bottom flange of a beam with the original length L_p has already been heated up to the bath temperature while the top, cold flange approximates the ambient temperature (the influence of heat conduction and co-effect of the web plate is disregarded). The rigidity of the beam resists the deformation caused by the bimetallic effect. The compressive stress in the bottom heated-up flange is in equilibrium with the compressive stress in the colder top flange. Under actual hot-dip galvanizing conditions, the value of this stress may reach values exceeding 500 MPa in a material with a sufficiently high yield point.

A change of the geometry of the part after its cooling depends on the rate of its plastic deformation. Beams made of steel of a lower grade achieve their yield point earlier and the major part of the thermal deformation occurs in the plastic

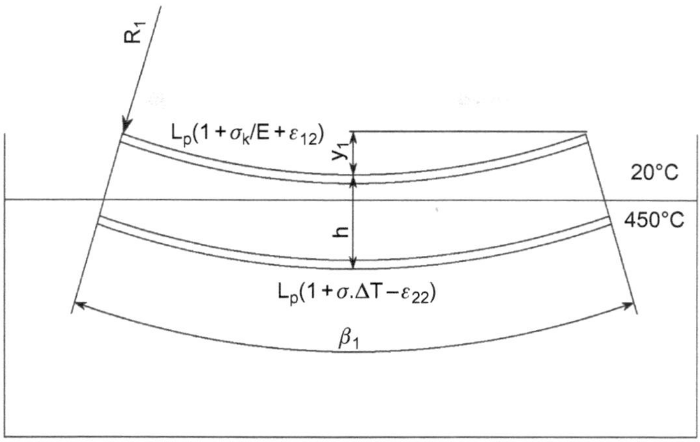

Figure 10.15 Bimetallic effect during immersion of a beam in zinc melts.

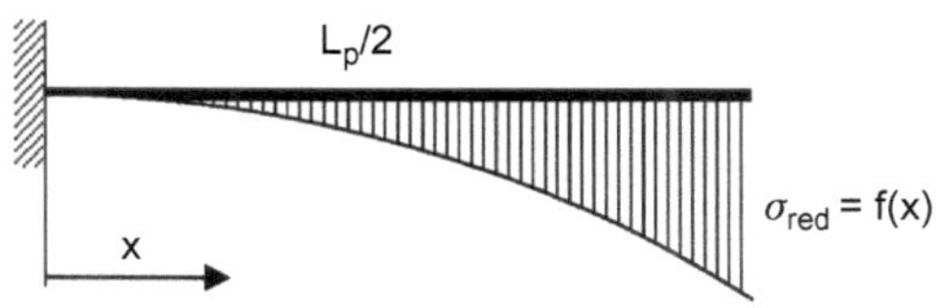

Figure 10.16 Reduced stress in the web plate of a beam immersed in the zinc bath up to a half of the web plate height (solution with an alternative fixed beam of half the length).

deformation range. However, for beams made of high-strength steel, deformation mostly occurs in the elastic range and the beams are affected by permanent deformation to a lesser extent.

With regard to limited allocation of deformation energy (the limitation is given by the thermal expansion of steel), the relative deformation in the outer edges of prismatic beams varies in fractions of the nominal values of ductility and the risk of achieving the limit condition leading to nucleation of cracks is low.

The reaction inside the beam acting against its deflection by the bimetallic effect is distributed by the web plate of the beam. The highest values of planar stress acting in the web plate are achieved at the end of the beam (Figure 10.16) in the place of contact of the web plate with the top flange. Besides tensile stress σ_t there is also shear stress τ_{sp} resulting from mutual action of the web plate and flange. Reduced stress in the web plate depends on the geometry of the beam and the regime of its immersion in the liquid metal and often achieves values at the yield point of steel. The relative deformation of the outer edges of the web plate (unlike the outer edges of the flanges) may be considerably higher. Thus, the area of the web plate at the end of a beam represents a structural element prone to nucleation of LMA cracks.

10.8.2 Gradual heating up

A steel article immersed in the zinc melt at the temperature of approximately 450°C gradually heats up from the surface to the core (Figure 10.17). The rate of heating is proportional to the rate of heat conduction by steel and the rate of heat transfer from the liquidus to the wall of the article. Both these parameters change with the instantaneous temperature, and the mathematical model describing all variable states during hot-dip galvanizing (design variants, various immersion regimes, etc.) would be very complex.

Also the stress caused in the steel by the temperature gradient during gradual heating from the surface to the core reaches, similarly to the bimetallic effect, the yield point of the material. The peaks of relative deformations caused this way are well below the material ductility values.

Evaluation of the stress state must respect the fact that steel is heated up at a certain rate (Figure 10.18). This fact can be used for proper setting of the conditions of immersion of a galvanized part in molten zinc. The expansion rate of a particular structural element corresponds to the heating rate of a steel article immersed in liquid metal.

From the point of view of permanent deformation you must also take into account that a structural element heated up to the liquid metal temperature and joined to a structural element that is still cold cannot be compacted immediately, but the rate of its compression is given by the creep rate of the used steel at the particular temperature. In spite of this you can assume that deformation caused by thermal stress will mainly occur in the bottom, heated up, and compacted element. At the same time the stress in the top element caused by tension and bending may reach values of the nominal yield point.

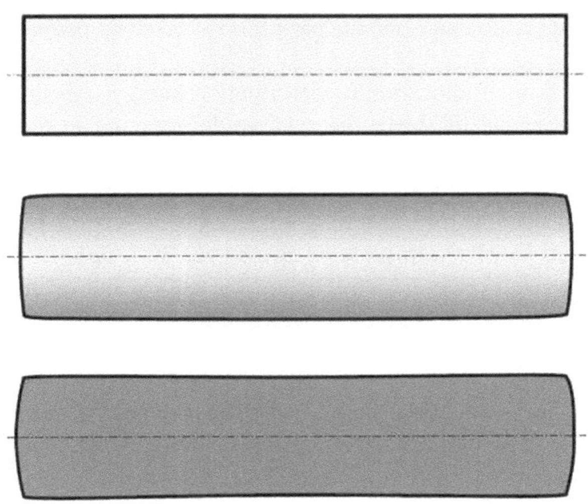

Figure 10.17 Gradual heating up of a cylindrical body immersed in molten metal.

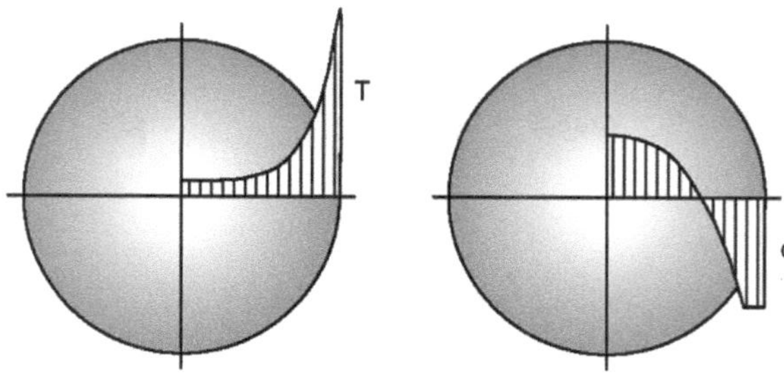

Figure 10.18 Temperature gradient T and axial tensile stress σ in the cross-section of a round bar immersed in liquid metal (shown for a certain random moment).

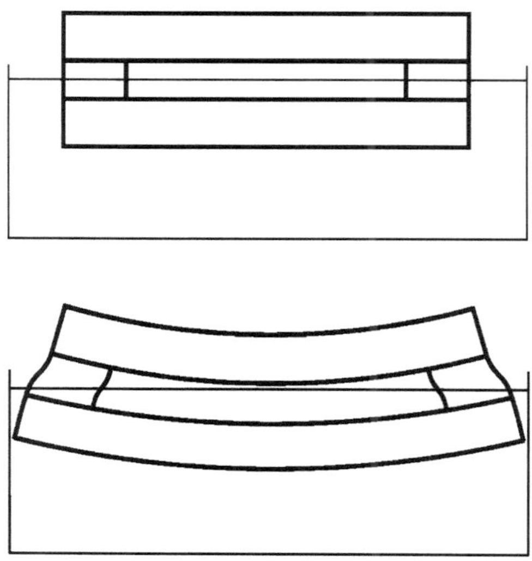

Figure 10.19 Diagram of deformation of a frame partly immersed in liquid metal.

10.8.3 Uneven heating of a frame structure

Figure 10.19 illustrates deformation of a frame the bottom stringer of which is immersed in liquid zinc. Depending on the cross-sectional area of both the stringers and rigidity of the structure, the opening force acting upon the two side arms of the frame may in standard practice vary on the order of 10^6 to 10^7 N. The rate of opening of the arms approximates the value of 5.10^{-2} m at a frame length of 10 m.

The bottom stringer of an unevenly heated frame as in Figure 10.19 changes its length depending on its immersion time in the zinc bath. The rod is gradually heated and expands, but at the same time it is subject to buckling stress due to the action of the cold top stringer. Thermal expansion proceeds relatively quickly and the stringer first achieves a relative deformation that is very close to the value corresponding to the thermal expansion of steel ($\varepsilon = \alpha \cdot \Delta T$). It is only after a

Figure 10.20 The risk of part damage by the action of variable thermal stress is increased by the use of hollow bracing elements with insufficiently sized drain holes.

certain delay that the rod starts to be compressed by the buckling axial force and its length is slightly reduced; if the yield point is exceeded, material compaction and permanent deformation of the structure may occur.

10.8.4 Three-phase *liquid metal assisted (LMA) crack nucleation model [51]*

To nucleate an LMA crack, a sufficient level of tensile stress must be achieved. In spite of this, experiments and practical experience show that on the one hand, on contact with liquid metal, steel articles are able to withstand significant plastic deformations without their integrity being disrupted regardless of the steel grade. On the other hand, the limit condition for brittle fracture may be achieved at a tensile stress below the nominal yield point values. In the trusses shown in Figure 10.20 affected by LMA cracks (Figure 10.21) the nucleation of cracks was not preceded by plastic deformation. With regard to a relatively small difference of the specific weights of steel and zinc in the last stage of immersion of hollow parts in the zinc melt, the immersion rate was considerably reduced. The air captured in the cavities of the parts caused a floating effect just at the moment when the bottom part of the top stringers came into contact with the liquid metal. This situation caused that the material of the stringer, which was first subject to significant tensile and bending stress from the expanding bottom stringer, was compacted in the part that got quickly in contact with liquid metal, and subsequently (after immersion of the whole top stringer and its heating up and expansion) the stress acting in the critical notch area changed to tensile stress again.

A detailed analysis of conditions under which material integrity disruption frequently occurs indicates that the critical stress concentrator was subject to variable stress before the occurrence of the crack, first to considerable tensile stress (usually

Figure 10.21 All parts on a suspension affected by LMA cracks.

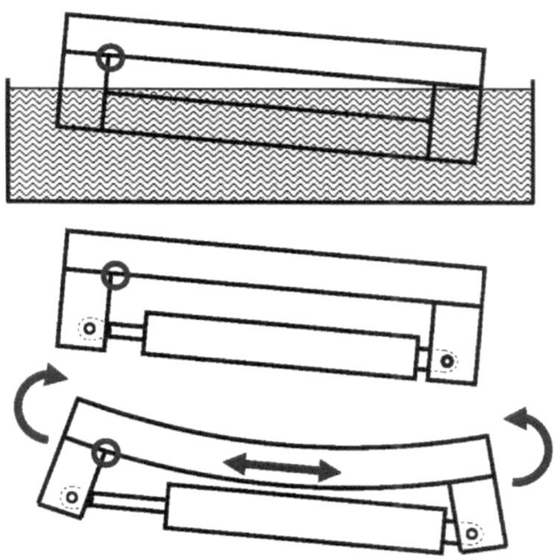

Figure 10.22 Phase I – tensile and bending stress.

combined with bending) and then to pressure associated with material compaction and then to tensile stress again.

Phase I

The critical detail is located in such a way on the part that during its immersion in liquid zinc the critical notch is first subject to considerable tensile stress when tensile force combined with bending torque acts upon the given element (Figure 10.22). Reduced stress in the article locally achieves the yield point of the steel and is associated with movement of dislocations inside the steel structure. In a

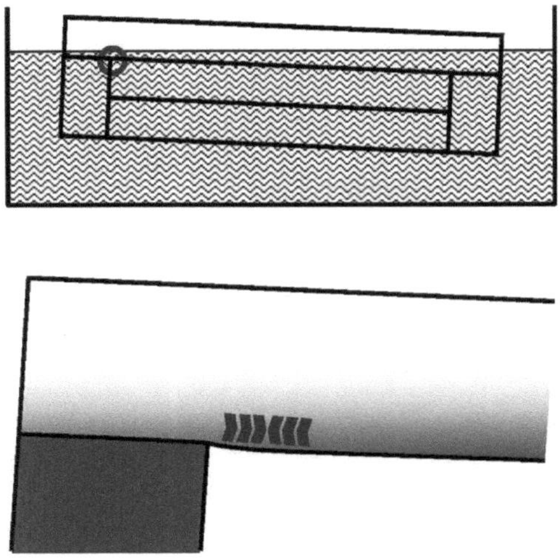

Figure 10.23 Phase II – material compaction after contact with liquid metal.

weld and the heat-affected zone of the weld, there are always a great number of imperfections where smaller resistance is offered to deformation energy than in the homogeneous structure of steel with a regular crystal lattice.

Phase II

As soon as the critical detail on the article comes in contact with liquid zinc (Figure 10.23), the material starts to gradually heat up at that point. However, its expansion is prevented by the still cold top part of the article, and the material in the critical detail, which was subject to extreme tensile and bending stress in the previous step, is compacted, which is accompanied by movement of dislocations inside the crystal lattice again.

Phase III

After a certain time delay, the remaining part of the top stringer comes in contact with the liquid zinc, which causes its expansion as a whole and extension of its length, while the critical detail, which was plastically deformed by the action of tensile and bending stress in the first step and subsequently compacted by compressive stress, is exposed to significant tensile stress again (Figure 10.24).

In case of uniaxial stress by simple tension of an article placed in a zinc bath, the yield point of steel may be reached without any fracture (Figure 10.27). In tests carried out, specimens exposed to variable stress with tensile and compressive history suffer from brittle fracture at quite a small relative deformation characterized by several parallel fracture surfaces (Figure 10.25). The fracture exhibits an intercrystalline character (Figure 10.26).

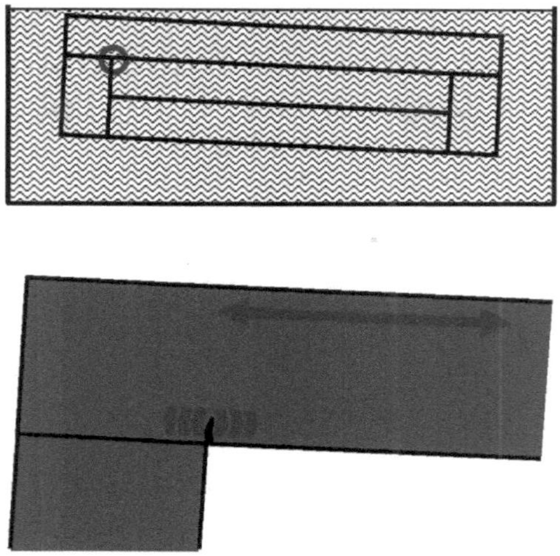

Figure 10.24 Phase III – after complete immersion of the frame in liquid metal the packed material of the notch is subjected to tensile stress from the expanding top stringer again and a brittle fracture occurs.

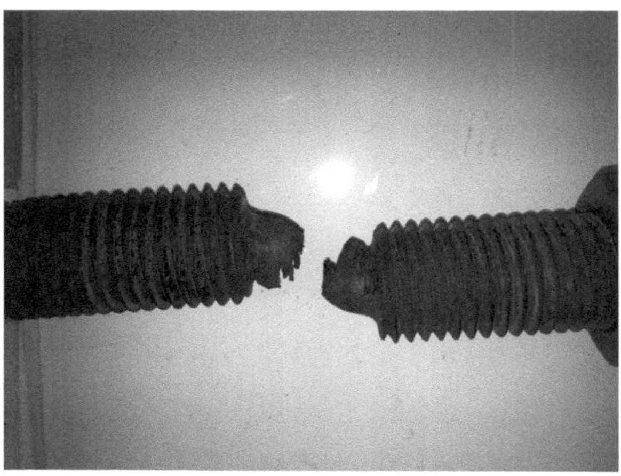

Figure 10.25 Character of the fracture after a simulated three-stage model.

Simulation of the three-phase model conditions shows that for disruption of integrity during hot-dip galvanizing, the condition of a combination of more unfavorable influences must be met. Stressing by simple tension on simultaneous contact of specimens of S355J2 and AISI 4130 steel grades with liquid zinc may not cause fracture even when the yield point has been achieved. The AISI sample without a notch (at the front in Figure 10.27) resists stress at the ultimate strength while the same specimen with a technological notch breaks at a very low stress even without the fracture being preceded by contact with liquid metal (Figure 10.28). The cause of

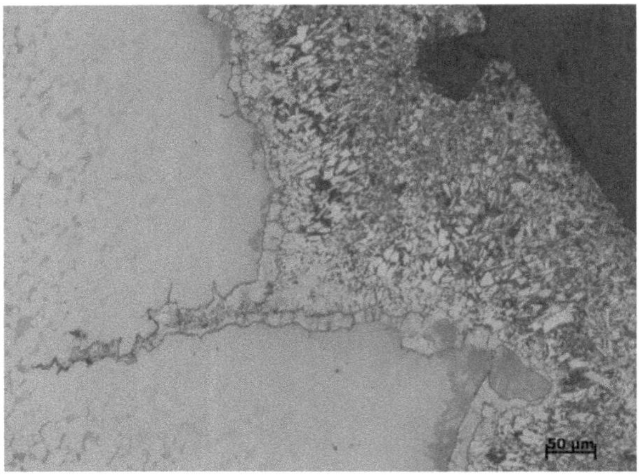

Figure 10.26 Soldered fracture after a simulated three-stage model.

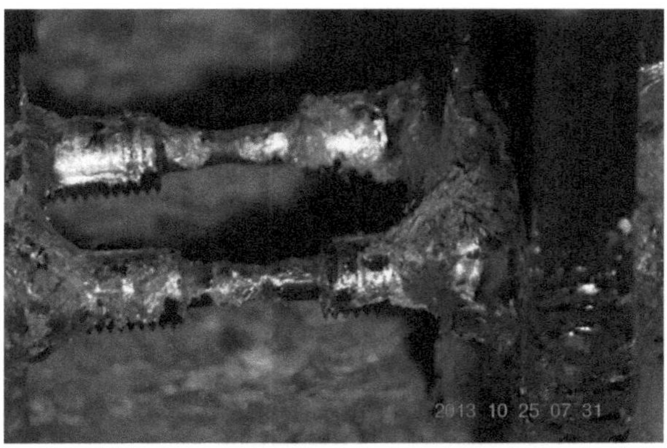

Figure 10.27 Specimens of the S355J2 and AISI 4130 steel grades immersed in a zinc bath subject to simple tension.

embrittlement is found in the heat-affected zone of the weld. If the proper welding procedure is not met, conditionally weldable steel always suffers from a number of imperfections from precipitates of nitrides, phosphides, carbides, or silicates through numerous dislocations to micro- and macro-cracks. On the other hand, steel grades with guaranteed weldability (with a very low carbon content) may carry stresses exceeding the yield point in spite of the presence of technological notches. Strangulation of both the specimens in Figure 10.29 (at the front the specimen of AISI 4130 steel, at the back specimen of S355J2 steel with a technological notch) proves that the ultimate strength of steel has been achieved. Supporting the hypothesis of a three-phase model is the fact that the separation of the specimen of S355J2 occurred during the third phase of the experiment outside the strangulation zone (Figure 10.30). The cause of cracks is neither the limit stress nor extremely large plastic deformation.

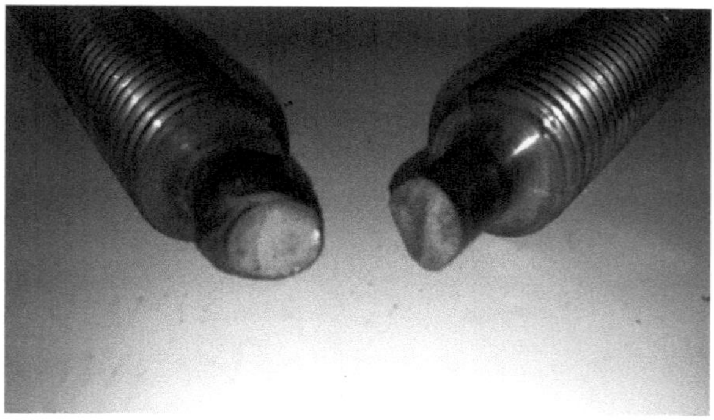

Figure 10.28 Appearance of the fracture surface after direct detachment of a specimen of the AISI 4130 steel with a technological notch.

Figure 10.29 Specimens stressed at ultimate strength before immersion (at the front a specimen of AISI 4130 steel, at the back a specimen of S355J2 steel).

Figure 10.30 Separation of the specimen of S355J2 occurred outside strangulation.

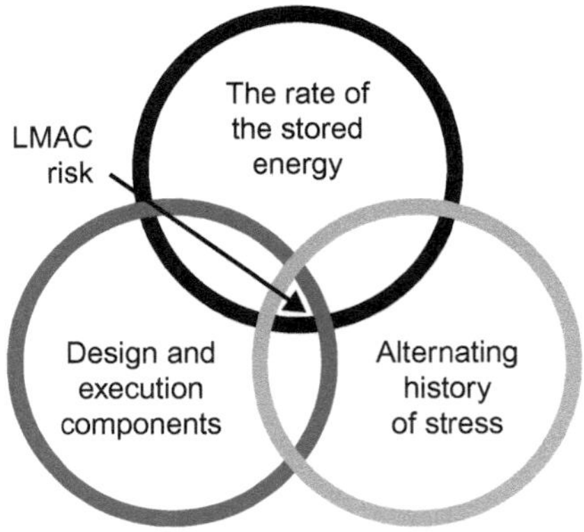

Figure 10.31 Convergence of adverse factors for LMAC of type "A."

Reaching the limit state as a result of stored elastic energy due to bimetallic thermal tension is one condition for the liquid metal assisted crack of type "A." LMAC of type "A" can occur as a result of the convergence of a number of adverse factors. The most significant negative factors are described by the diagram in Figure 10.31. To a large extent these factors can be controlled:

- rate stored elastic energy induced of bimetallic temperature tension
- variable thermal stress history with alternating tensile and compressive stress
- inappropriate design and/or poor execution of the part

The initial tension stage (Phase I) is not necessary for nucleation of an LMA crack. Parts designed to be made of materials with a significantly different section thickness may be affected by LMA cracks of type "B" [51]. During typical immersion and heating up, the massive element prevents free expansion of the thin-walled element with faster heating up, which gets compacted (Figure 10.32). The massive element is heated up and expands with a time lag, causing tensile stress in the thin-walled element. The critical location for occurrence of LMA cracks is the weld toe. A crack in structural steel is characterized by intergranular fracture disruption (Figure 10.33).

LMA cracks of type B are only formed in the course of cooling of the galvanized part after the application of the coating. The occurrence of the crack is caused by significantly different thicknesses of mutually coupled elements. An element with a thick section heats up more slowly than an element with a thin section. After immersion in the zinc bath the thin-section element is first compacted in two stages – first during its heating up, when it cannot expand freely due to the presence of the thick-section element, and then, after sufficient heating up, it is exposed to the action of the coupled thick-section element, which is heated up and expands with a certain time lag. After emerging from the zinc bath the thin-section element stressed and compacted this way cools more quickly than the thick-section element. Subsequent shrinking of the thick-section element causes considerable tensile stress in the joint with the thin-section

Figure 10.32 Construction part combining elements with a different wall thickness [23].

Figure 10.33 Construction part combining elements with a different wall thickness.

element. On achievement of the limit state, cracks may occur (Figure 10.34). Contributing negative factors are non-conforming welding and harmful influence of hydrogen. Possible preventive measure during hot-dip galvanizing is sufficient preheating of the components.

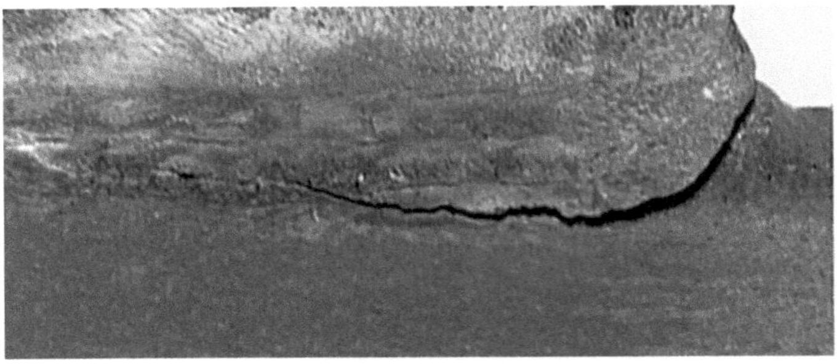

Figure 10.34 LMA crack of type B occurs during cooling after galvanizing (12x magnification) [51].

10.8.5 Cracks with their origin in a weld

The results of the experiments described in the previous Section 10.8.4 indicate that, besides LMA cracks, galvanized parts may be affected by cracks originating in a weld (see also Section 10.5). Figure 10.6 shows a crack originating in a weld where variable thermal stress during hot-dip galvanizing caused interconnection of already existing discontinuities – dislocations and micro-cracks (or macro-cracks) in the weld metal resulting from shrinking during the welding.

Every weld is, regardless of the quality of its workmanship, a significant stress concentrator. At the boundaries of the weld metal phosphides, nitrides, silicates, and carbides precipitate, which reduce cohesion; the material of the weld metal has a different composition and structure from the base-joined metal. An abrupt change of material characteristics is also typical for welds. Grains in the welding metal and the heat-affected area become coarse and there is always considerable spatial stress causing the movement of dislocations and occurrence of numerous micro-cracks.

The fact that steel structures made of structural ferritic-pearlitic steels of higher quality grades are more frequently affected by cracks during hot-dip galvanizing is related to welding to a considerable extent since higher strength of steel is achieved through a higher content of carbon, which has a negative impact on the quality of welded joints. Guaranteed weldability of these steel grades is conditional on the fulfillment of certain conditions for a proper welding procedure. These steel grades are more sensitive to non-observance of the technological discipline of welding than structural steel of the usual grades characterized by relatively low carbon content. Welded joints of such structures are affected by imperfections to a higher extent.

10.8.6 Influence of tin on the LMAC risk

When analyzing the LMAC phenomenon some authors found out that samples taken from parts affected by galvanizing cracks exhibited a relatively high concentration of some low melting point metals that are added as alloying additions to the zinc melt, in deep locations, near the crack tip. This led to the assumption that atoms of

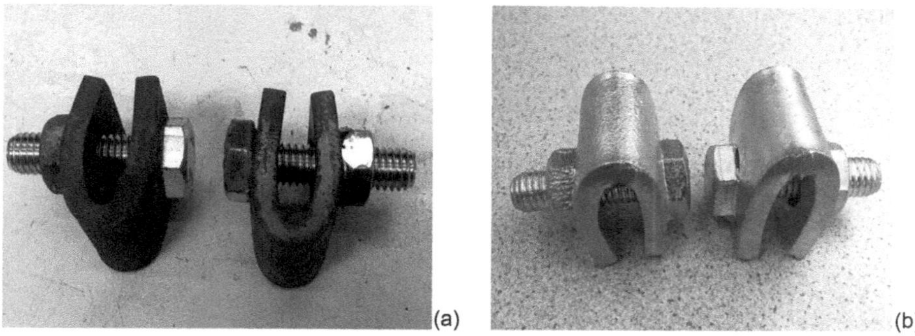

Figure 10.35 "V" specimens: a) ready for the experiment, b) after galvanizing.

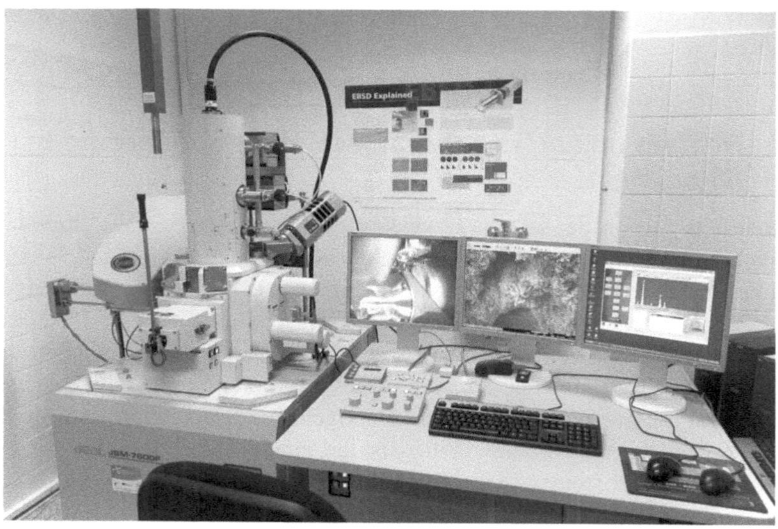

Figure 10.36 JEOL JSM-7600F electron scanning microscope [52].

metals with relatively high affinity to steel are able to penetrate into it along grain boundaries, causing intercrystalline separation (decohesion) of the material.

Experiments conducted with "V" specimens made of statically pre-stressed S235JR and S355J2 steels just before exposure in a zinc bath with added tin do not prove this theory (Figure 10.35). An SEM analysis with an electron scanning microscope (Figure 10.36) JEOL JSM-7600F with an FEG field, with the resolution of up to 0.8 nm and magnification of 1,000,000, shows that after a usual immersion time of the specimen in the galvanizing bath, the concentration of tin at the boundaries of steel grains in its outer surface layer (Figure 10.37) to the depth of 1 μm is at the detection limit of the measuring apparatus. At a depth of more than 1 μm the tin content is not measurable.

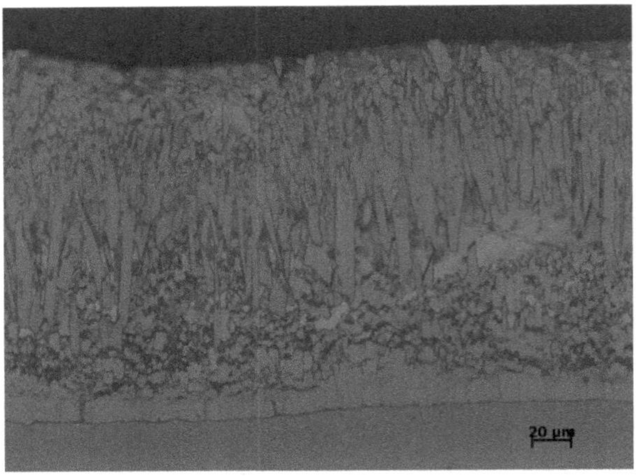

Figure 10.37 Metallography of a "V" specimen galvanized in a galvanizing bath with the tin content of 2%.

10.8.7 Heating up and expansion rate of steel immersed in liquid metal

The expansion rate of a structural steel element immersed in a liquid metal corresponds to its heating rate. As the instantaneous temperature of an element that is being heated up has a significant gradient (Figure 10.18), the instantaneous length of a rod is proportional to its *reference temperature* (the surface of the rod has a temperature close to the liquid metal temperature while the temperature in its core is relatively low). The reference temperature of a rod is the temperature that would cause the same relative elongation in the homogeneous temperature field in the rod as is achieved in the temperature field with the gradient corresponding to gradual heating up of the rod from its surface to the core. The heating up rate of a steel body immersed in molten zinc has a key influence on the variable temperature stress caused in it.

It is a well-known fact that steel structures made of typical grade structural steel (with a low nominal yield point value) are less prone to LMA cracks while parts made of high-strength steel are characterized by higher susceptibility. The cause of this higher susceptibility is mainly the fact that a higher yield point is a precondition for achieving a higher level of variable temperature stress. Two identical structural parts made of different steel grades are exposed to different temperature stress levels under the same hot-dip galvanizing conditions depending on the yield point of the used steel. The material of a part made of a lower steel grade relaxes on achieving the yield point while in a part made of high-strength steel the stress may keep rising depending on the temperature difference between the cold and heated section of the part. This difference of *reference temperatures* depends on the rate of immersion of the part in the liquid metal. To reduce the risk, galvanizing conditions must ensure that the amount of stress caused in the structure by its expansion should not reach critical values.

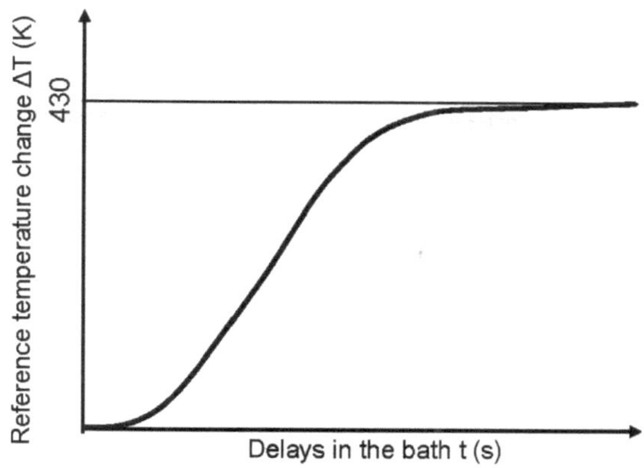

Figure 10.38 Chart of time evolution of relative deformation of a prismatic steel rod immersed in a zinc bath.

The time evolution of relative deformation of a prismatic steel rod immersed in a zinc bath is shown in the chart in Figure 10.38 [51]. Knowledge of the heating up profile of rods of various cross-sections can be used to reduce the risk of LMAC related to hot-dip galvanizing. Thin elements of the structure are heated up and expand more quickly than massive elements.

If the elongation of a rod corresponds to a certain change of its reference temperature, for a frame structure the peak value of variable temperature stress is proportional to the difference of reference temperatures between the bottom and top stringers. Such a difference can be determined for the conditions of the process of immersion of a frame structure in liquid metal that certain critical stress should not be exceeded. Practical experience shows that LMA cracks rarely occur in parts made of S235 steel. On the other hand, the use of steel grades over S420 represents a higher risk. If the reference stress in the part should not exceed the value $\sigma_v = 235$ MPa, the difference ΔT between the coldest and hottest part of the galvanized part must not be higher than 186 K. Immersion elements in the zinc bath must be done in the time interval t_c.

The curves in Figure 10.39 show the evolution of heating up of the top and bottom stringer of the frame structure. A properly adjusted time shift t_c (time of passage of the frame cross-section through the zinc bath surface) should guarantee that the required difference of reference temperatures will not be exceeded to make sure that the bimetallic temperature tension will not be higher than 235 MPa.

Measurement of relative deformation is shown in Figure 10.40. Different action of heat on a beam immersed flat in a zinc bath up to the half of the web plate height caused its deflection. Changes of the beam geometry over time were registered by shifts of points on arms fixed to the beam on the vertical plane, perpendicularly to its longitudinal axis.

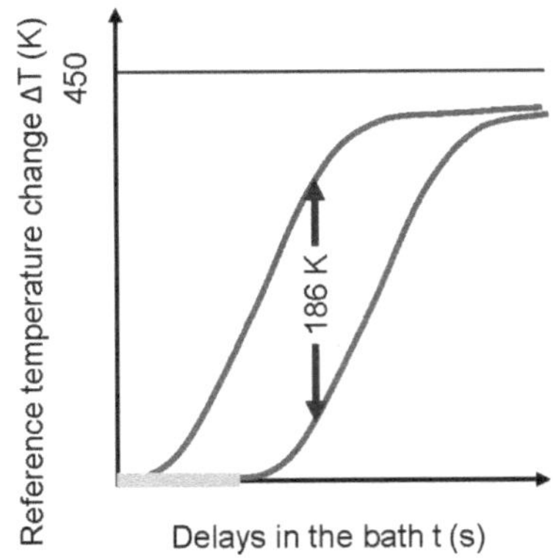

Figure 10.39 Critical interval of passing of the frame structure through the zinc bath surface marked gray.

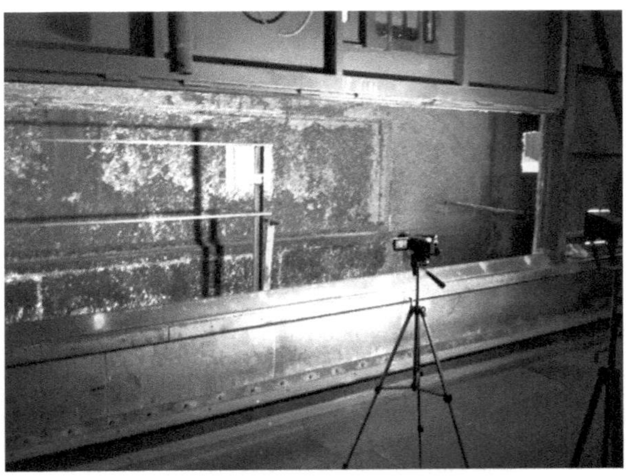

Figure 10.40 Measurement of the time evolution of deformation of a beam immersed in liquid metal.

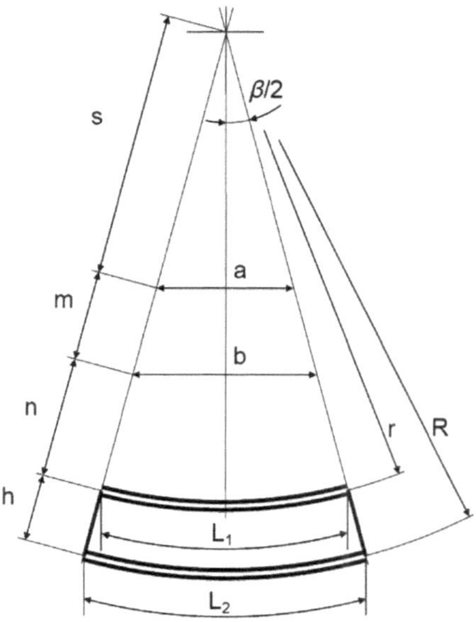

Figure 10.41 Analysis of the conversion of the shifts of the points on the arms to length changes of both the beam flanges.

Rods with various section thicknesses have heating up curves with various gradients. Shifts of the points on the fixture arms can be converted to length changes of both the beam flanges (Figure 10.41):

$$\sin\frac{\beta}{2} = \frac{b-a}{2m}$$

$$\beta = 2\ \text{arcsin}\frac{\beta}{2}$$

$$s = \frac{a}{2\ \sin\frac{\beta}{2}}$$

$$r = s + m + n$$

$$R = r + h$$

$$L_1 = r \cdot \beta$$

$$L_2 = R \cdot \beta$$

Standardization

11

To ensure the unification of technical requirements for hot-dip galvanized coatings and methods of evaluation of conformity, hot-dip galvanizing is governed by technical standards in a similar way to other branches. For member states of CEN (European Committee for Standardization) European technical standards are valid. These are based on common experience as well as research results and they focus on achieving optimum benefits for the community. On their basis, by complete adoption of requirements set forth by the European standard (that as a common European standard has been established in accordance with the European Community legislation by mutual agreement of national standards bodies), common national technical standards are introduced.

11.1 Basic standards for hot-dip galvanizing

EN ISO 1461 Hot-dip galvanized coatings on fabricated iron and steel articles – specification and test methods

This technical standard contains specifications for properties of and methods of testing for galvanized coatings. It is a document governing the execution of hot-dip galvanized coatings. It contains specifications of general requirements that should be agreed and fulfilled between the purchaser and the galvanizer. In terms of coating quality assessment, the acceptance tests provisions of Chapter 6, which contains Section 6.1 Appearance and 6.2 Coating Thickness, are binding for both the parties. A selection method for inspection samples and a process of unified evaluation of measured parameters are prescribed for acceptance tests.

EN ISO 14713-1 Zinc coatings – guidelines and recommendations for the protection against corrosion of iron and steel in structures. Part 1: general principles of design and corrosion resistance

Part 1 deals with general design principles and contains information about zinc corrosion in various environments and more information about the selection of an anti-corrosion system and its maintenance.

EN ISO 14713-2 Zinc coatings – guidelines and recommendations for the protection against corrosion of iron and steel in structures. Part 2: hot-dip galvanizing

Part 2 contains instruction for correct structural design and execution of products designed for hot-dip galvanizing and for selection of suitable materials. It draws attention to specific conditions of hot-dip galvanizing and to specific aspects that must be taken into account for application of this anti-corrosion system.

Hot-Dip Galvanizing of Steel Structures. DOI: http://dx.doi.org/10.1016/B978-0-08-100753-2.00011-2

EN ISO 14713-3 Zinc coatings – guidelines and recommendations for the protection against corrosion of iron and steel in structures. Part 3: sherardizing

Part 3 deals with sherardizing, which is a thermally applied diffusion coating.

EN 13811 Sherardizing – zinc diffusion coatings on ferrous products – specification

This technical standard divides coatings produced by sherardizing into three classes. It specifies minimum requirements for zinc dust used for sherardizing and the requirements for minimum coating thicknesses. This standard will soon be replaced by EN ISO 17688.

EN ISO 10684 Fasteners – hot-dip galvanized coatings

This technical standard specifies material, process, dimensional, and some performance requirements for hot-dip centrifuged galvanized coatings applied to threaded steel fasteners from M8 up to M64 made of steel of the strength class up to 10.9 (for bolts and screws) and 12 (for nuts).

11.2 Technical standards related to hot-dip galvanizing

EN ISO 2063 Thermal spraying – metallic and other inorganic coatings – zinc, aluminum and their alloys.

This standard relates to metallic coatings of zinc and aluminum or their alloys applied by thermal spraying and designed to protect steel against corrosion. It deals with their characteristics and contains a classification and identification of these coatings in relation to their thickness and methods of their testing. The standard further deals with surface preparation, application of coatings, and their characteristics.

EN 1179 Zinc and zinc alloys – primary zinc

This European standard specifies a classification, chemical composition, marking, and other requirements for primary zinc. Zinc classes included in this standard are internationally tradable.

11.3 Tests

Test methods are closely related to zinc coating quality assessment and evaluation. EN ISO 1461 contains information for specifying the methods of zinc coating measurement that can be used and under what conditions. In cases of dispute it states that the thickness determined using the gravimetric method in accordance with EN ISO 1460 is decisive. For coating adhesion the standard does not specify any method for such a test. If based on a special agreement between the purchaser and galvanizer, an adhesion test should be conducted, the agreement should also contain provisions about the test method and responsibility for its execution.

EN ISO 1460 Metallic coatings – hot-dip galvanized coatings on ferrous materials – gravimetric determination of the mass per unit area

In accordance with this method, the coating thickness is determined by calculation from the weight of coating removed from a known area of the surface of a cut-off sample. It is a destructive method, which also involves high demands for apparatus.

EN ISO 1463 Metallic and oxide coatings – measurement of coating thickness – microscopical method

Another coating thickness measurement method is the microscopical method, the use of which requires a sample to be cut from the part under test, i.e., its destruction. The coating thickness is determined using the microscope scale on a metallographic sample. This method is also demanding in terms of apparatus.

EN ISO 2177 Metallic coatings – measurement of coating thickness – coulometric method by anodic dissolution

The coulometric method is based on determining the volume of a metallic coating removed from a surface with a known area by conversion of the charge necessary for its galvanic dissolution. This method is conditionally destructive; the tested place can be subsequently repaired in accordance with EN ISO 1461 if the test can be done without cutting off the sample.

EN ISO 2178 Measurement of coating thickness – magnetic method

Although this standard is called the "magnetic method," it actually covers two different methods.

The first method is based on measuring the attractive force acting on a permanent magnet. The measurement is carried out using a pencil thickness gage the calibration of which is based on the physical principle that the attractive force is indirectly proportional to the thickness of the non-magnetic layer between the magnet and ferromagnetic substrate. The magnet is pushed away from the measured surface by a spring and at the moment of tearing off, the spring compression value is recorded. This method is very inaccurate and it does not provide data that could be evaluated in accordance with EN ISO 1461. It only provides approximate information regarding whether the coating thickness is acceptable.

A much more accurate method is a magnetic method based on the measurement principle of reluctance during the passage of magnetic induction flux through the coating and the ferromagnetic basis metal. It does not involve high demands for apparatus and its great benefit is its flexibility. The measurement immediately provides results that can be quickly processed and evaluated. If the principles of a proper measurement process and evaluation of the measured values are observed, it provides reproducible results that are comparable to the measurement results achieved using the gravimetric method. This is the most frequently used method of measurement of the thickness of a non-magnetic coating on a ferromagnetic substrate.

EN ISO 2360 Non-conductive coatings on non-magnetic electrically conductive basis materials – measurement of coating thickness – amplitude-sensitive eddy-current method

This method is not recommended for zinc coating thickness measurement, but as a non-destructive method it is suitable for determination of the thickness of an organic coating on the zinc coating in duplex systems. This standard describes a method of non-destructive measurement of the thickness of non-conductive coatings on non-magnetic electrically conductive (usually metallic) base materials with devices using the amplitude-sensitive eddy-current method.

Note

This method can also be used for measurement of non-magnetic metallic coatings on non-conductive basis materials.

The method can be especially used for thickness measurement of most coatings produced by anodic oxidation, but it is not suitable for all conversion coatings as some are too thin for measurement using this method (see Chapter 7).

EN ISO 4624 Paints and varnishes – pull-off test for adhesion

This standard describes methods for determining the adhesion by carrying out a pull-off test on a single coating or a multi-coat system by measuring the tensile stress required to detach or pull of the coating on an exactly delimited area in the perpendicular direction to the substrate.

EN ISO 16276-1 Corrosion protection of steel structures by protective paint systems – assessment of, and acceptance criteria for, the adhesion/cohesion (fracture strength) of a coating – Part 1: pull-off testing

The standard specifies procedures for assessing the fracture strength of a protective paint coating of any thickness on a steel substrate of thickness not less than 10 mm. The procedures given in this part of ISO 16276 are based on methods used with different types of pull-off test equipment.

EN 45014 General criteria for supplier's declaration of conformity

The standard specifies general criteria for the supplier's declaration of conformity when it is desirable or necessary that conformity of a product, process, or service be expressed regardless of the concerned branch. In such cases the supplier can declare conformity to normative documents at their responsibility.

11.4 Corrosivity of atmospheres

EN ISO 12944-2 Paints and varnishes – corrosion protection of steel structures by protective paint systems – Part 2: external environment classification

This important standard belongs to the series of technical standards for paints and varnishes, but it contains a classification of the external environment in relation not only to organic painting materials but also to selected metallic coatings and serves as an important tool for designing anti-corrosion systems.

EN ISO 9223 Corrosion of metals and alloys – corrosivity of atmospheres – classification, determination and estimation

The standard defines degrees of corrosivity and provides information for estimating the degrees of corrosivity based on the knowledge of the local environment.

EN ISO 9224 Corrosion of metals and alloys – corrosivity of atmospheres – guiding values for the corrosivity categories

The standard specifies guiding values of corrosion losses for metals and alloys exposed to outdoor natural atmospheres for a period longer than one year. This standard should be used together with EN ISO 9223.

EN ISO 9225 Corrosion of metals and alloys – corrosivity of atmospheres – measurement of environmental parameters affecting corrosivity of atmospheres

This standard specifies measurement methods of parameters necessary for the estimate of corrosivity used in EN ISO 9223 for classification of corrosivity of atmospheres and measurement methods of environmental parameters for standardized estimate or corrosivity on the basis of calculated corrosion rates of basic metals in the first year of exposure and for informative estimate of corrosivity based on the environmental characteristics of the place of exposure.

EN ISO 9226 Corrosion of metals and alloys – corrosivity of atmospheres – determination of corrosion rate of standard specimens for the evaluation of corrosivity

This standard specifies methods that can be used to determine the corrosion rate with the use of standard specimens. The values obtained by measurement (corrosion rate after the first year of exposure) are used as the classification criteria for determination of corrosivity of atmosphere in accordance with EN ISO 9223. They can also be used for informative evaluation of corrosivity of atmosphere outside the framework of EN ISO 9223.

11.5 Other selected standards and regulations

EN 10025-2 Hot rolled products of structural steels – Part 2: technical delivery conditions for non-alloy structural steels

A great majority of parts designed for hot-dip galvanizing are made of hot rolled structural steel. The standard contains an option to order steel with defined chemical composition for hot-dip galvanizing in terms of the chemical composition of steel. The standard defines three classes as defined by their silicon and phosphorous content.

EN 1090-2 Execution of steel structures and aluminum structures – Part 2: technical requirements for steel structures

Annex F of the standard dealing with anti-corrosion protection contains brief principles for hot-dip galvanizing from the point of view of coating specification and integrity of hot-dip galvanized steel structures.

11.6 Management systems

The topics of introduction of quality management or environmental management systems are covered by the respective European standards.

EN ISO 9001 Quality management systems – requirements

This standard specifies requirements for the quality management system in cases where an organization needs to prove its ability to permanently guarantee the conventional quality standard, including processes for its continuous improvement.

EN ISO 14001 Environmental management systems – requirements with an instruction manual

The standard specifies requirements for the environmental management to help an organization create and introduce a policy and define goals that will comprise requirements of legal regulations and other requirements that are relevant for the organization and information about important environmental aspects.

OHSAS 18001 Occupational health and safety management systems – requirements

The standard specifies requirements for assessment of occupational health and safety. The OHSAS standard will be canceled after publication of its content by incorporation in an international standard by issuing of a separate international standard.

Legislation

12

This chapter dealing with legislation focuses on the environmental aspects related to the operation of hot-dip galvanizing plants. This chapter does not cite any particular legal regulations as they are subject to relatively frequent changes. The content of the chapter is limited to general information about the project approval process and operation of hot-dip galvanizing plants.

12.1 Evaluation of environmental influences

New investment intentions concerning facilities whose operation may affect the environment are subject to the EIA process (Environmental Impact Assessment) before implementation. The purpose of the Environmental Impact Assessment is to obtain an independent document for the decision-making process or measures and to contribute to sustainable development of the community. Impacts on public health and impacts on the environment are assessed, including influence on animals and plants, ecosystems, soil, rock environment, water, atmosphere, climate and landscape, natural resources, material property and cultural monuments, and their mutual influences and relationships.

12.2 Integrated permitting

Within the IPPC (Integrated Pollution Prevention and Control) process an integrated permit is issued for a specified facility whose operation may give rise to pollution of the environment. The permit defines binding conditions of operation so that a high standard of environmental protection can be achieved in line with the legislation of the European Union. The administrative proceedings concerning the issue of the integrated permit comprise verification whether the investor has selected Best Available Techniques (BAT) for the implementation of the project.

Such production processes and equipment are considered as BAT that ensure achievement of the best environmental protection parameters at acceptable economical costs.

12.3 Reference documents about the Best Available Techniques

To unify requirements for techniques and parameters for equipment with possible negative impacts on the environment, information about the Best Available

Hot-Dip Galvanizing of Steel Structures. DOI: http://dx.doi.org/10.1016/B978-0-08-100753-2.00012-4

Techniques (BAT) has been summarized by the European Union in so-called BAT Reference Documents (BREF). Hot-dip galvanizing is covered by the reference document BREF 14 – Part C Batch Galvanizing.

12.4 Waste

For the protection of soil and water there are rules for prevention of generation of waste and its handling. The legal regulation defines the right and obligations of persons in waste management and competences of public administration authorities. A waste catalogue has been compiled and handling of dangerous waste is subject to especially strict rules. A governmental authority checks the way operators handle their produced waste.

12.5 Atmosphere

Industrial hot-dip galvanizing involves production of particulate emissions to air. There is a general principle that operators of industrial plants must not release any pollutant into the atmosphere that might have a negative impact on the environment. A legal regulation defines maximum acceptable concentrations of selected pollutants in the air emissions released from the plant. The decision of issuing the integrated permit in accordance with IPPC may set forth stricter conditions than required by the legislation depending on the local distribution of other sources of atmospheric pollution. Compliance with emission limits is documented by regularly conducted authorized measurements.

12.6 Land

Operation of chemical pre-treatment in galvanizing plants is associated with the risk of leakage of hazardous substances into the surrounding soil. In the EU countries there is an obligation of the investor to submit a baseline report for granting permission to build a new operation. The basic report may draw only on an authorized geologist and be based on a site survey and an evaluation of historical contamination (state of soil pollution) prior to project implementation. In the event of termination of operations, the facility operator shall return the grounds into a state corresponding to the baseline report.

Hot-dip galvanizing and the environment

13

Zinc as metal is a natural element and therefore its use for protection of steel against corrosion is quite natural, too. Its presence in the environment does not necessarily produce a disturbing effect (Figure 13.1). On the contrary, in a suitable combination with other natural materials as wood, stone, or glass an impressive aesthetic effect may be achieved (Figure 13.2).

Hot-dip galvanizing plants should be properly managed with regard to production of waste and emissions to air. In hot-dip galvanizing plants, pickling is generally based on hydrochloric acid solutions, so the description of the environmental impacts in this chapter uses this as its starting point. Modern hot-dip galvanizing plants are designed to meet strict legislative requirements for environmental protection with a large margin of safety. Operations using chemicals with hazardous characteristics are carried out in operating units separated from the working environment and are equipped with efficient pollutant filters.

13.1 Degreasing solutions

In the EU, some waste from degreasing falls under the dangerous waste catalog number 11 01 13. In the case of an alkaline degreasing agent, the whole bath saturated with grease may be required to be disposed of by an authorized contractual partner. For acidic agents, the waste has the character of solid sediment that is collected from the tank bottom when the solution has been pumped out. The volume of this waste is relatively small and it can be disposed of without any further treatment.

13.2 Pickling solutions

Ferrous chloride is generated in the pickling solution (when it is based on hydrochloric acid) during its gradual saturation with iron from steel parts pre-treated for galvanizing. A new pickling bath is started with the free acid content of 120 to 150 g/l and with a ferrous chloride content of about 50 g/l. During the useful life of the pickling solution, the free acid is gradually used up while the bath becomes saturated with iron. After reaching the saturation limit, no further pickling is possible and the bath may be handed over to an authorized entity for disposal as pickling waste.

Hot-Dip Galvanizing of Steel Structures. DOI: http://dx.doi.org/10.1016/B978-0-08-100753-2.00013-6

Figure 13.1 Hot-dip galvanized parts in a natural setting.

Figure 13.2 Hot-dip galvanized zinc coatings in architecture [54].

Saturated solutions of ferrous chloride can be used in wastewater treatment plants for separation of phosphorus. The action of bivalent or trivalent iron salts converts phosphorus to a solid, insoluble phase that precipitates in the sedimentation tank and may be landfilled after separation.

13.3 Stripping

Separated stripping baths are used to remove old zinc coatings or to remove zinc from process hooks/jigs or from rejected galvanized parts. During stripping, the content of iron in the solution must be kept to the minimum as iron reduces the yield of zinc from the solution for its further processing. As zinc reacts relatively strongly with hydrochloric acid while iron shows a higher resistance to the acid, the content of dissolved iron in the stripping bath can be eliminated by simple supervision of the operator, who must intervene at the moment when all zinc from the surface of the parts has been dissolved and effervescence of the bath has ceased.

A saturated solution of zinc chloride is an important raw material from which zinc can be separated successfully for further use in the chemical or pharmaceutical industry. Solutions from a well-managed stripping bath saturated with zinc and with a minimal content of iron are hazardous from the point of view of classification, equally to the pickling bath saturated with ferrous chloride, but it can be sold to an authorized company as raw material for further processing. Zinc chloride is also one of the constituents of flux and so it can be used for the preparation of flux in the produced form without any requirements for treatment.

13.4 Zinc ashes

During immersion of galvanized parts in the zinc bath, ashes with a high content of zinc are produced by oxidation of the zinc at the bath surface and from combustion of the applied flux, which are then present on the bath surface. Before the removal of the galvanized batch the ashes are skimmed to the end of the galvanizing bath with paddles. From there they are collected and stored as the other residues under the EWC catalogue number 11 05 02 so that they can be sold as a valuable raw material for further processing. The content of pure zinc in zinc ashes can reach approximately 70% and zinc can be relatively easily recycled by melting in a rotary furnace.

13.5 Skimmings

In hot-dip galvanizing plants that use the wet process, decontaminated parts are immersed in the zinc bath through a sal-ammoniac layer. In the course of time this will deteriorate. Deteriorated sal-ammoniac is periodically removed from the zinc bath and collected as skimmings with the EWC catalogue number 10 05 11 and is subsequently handed over to an authorized contractual company for disposal or recycling.

13.6 Filter dusts from air pollution control equipment

Particulates emissions from the galvanizing bath operation, mainly from flux combustion, are separated in high-efficiency fabric filters. This fine dust, which contains mostly sal-ammoniac, is a solid air pollutant and its concentration in the released air matter must not exceed the limit prescribed by the legislation. Captured filter dusts are collected as dangerous waste with the catalog number 11 05 03 and can be sent to a flux manufacturer for external recycling.

13.7 Dross

During the metallurgical diffusion of the galvanizing reaction that forms the coating, iron is released from the substrate and combines with zinc in the zinc melt in the ratio of 1:13, producing so-called dross (sometimes referred to as "hard zinc"), or iron-zinc alloy phase ζ. Dross has a higher specific weight than pure zinc and therefore it drops to the bottom of the galvanizing bath from where it is periodically collected. This waste has a high zinc content and is categorized under the EWC catalogue number 11 05 01. In the same way as zinc ashes, it is sold as an important raw material for processing in the chemical industry where zinc is recycled from it.

13.8 HCl emissions

The most significant air pollutant produced during the chemical pre-treatment of parts to be hot-dip galvanized is HCl emissions above the pickling baths. Therefore, in modern hot-dip galvanizing plants the chemical pre-treatment workplaces are separated from the working environment by using enclosures that have a high resistance to these chemicals and inside of which these emissions can be captured and separated. There are agents added to pickling baths that can be used to partly eliminate HCl evaporation. The pickling bath should not be operated at elevated temperatures to increase the pickling efficiency. The efficiency of the baths can be adequately controlled by proper management of the baths and timely adjustment of the chemical composition so that they can be operated in the ranges of optimum pickling times.

13.9 Heat recovery

The temperature of flue gas from galvanizing bath heating is used by means of efficient heat exchangers elsewhere in the production process for heating of hot service water, drying of the batch after flux application, or for supplementary heating of the degreasing or flux baths to ensure their required efficiency. Thus, besides reducing the energy demands of the process and the resulting cost savings, the impact of hot-dip galvanizing technology on the environment is also reduced.

13.10 Low water consumption

Proper management of the pre-treatment baths also leads to a considerable reduction or elimination of water consumption as rinsing water can be completely reused to replenish losses in the pickling baths instead of service water.

Sustainable development

14

Motto: *Mankind has the ability to make development sustainable – to make sure that the needs of the present are met without compromising the ability of future generations to meet their own needs* (Brundtland's definition).

Any materials, no matter how environment-friendly they are, influence the environment in some way. The aim of sustainable development is to do everything possible to limit the negative influences of human activities as much as possible so that the situation should not get any worse [54].

An example of a correct approach to the issue of sustainable development may be a forest manager, who, although his mission is timber production, can coordinate his activities in such a way that the volume of gradually felled trees in the course of time does not exceed the volume of newly planted ones, which means that the quantity of trees is not gradually reduced.

Mining and utilization of zinc can be approached in a similar way. The system of anti-corrosion protection of steel with zinc coating cannot be eliminated from our lives given its huge economic significance, but hot-dip galvanizing could be perceived as an activity that may be acceptable if the main conditions are strictly met:

- Renewable resources should be not consumed faster than the rate at which they manage to recover naturally.
- Exhaustible resources should not be consumed faster than the rate at which their replacement will be provided and to which a smooth transition will be possible.

  The volumes of produced pollutants must not exceed the capacity of the receiving environment.
- A part of current profits should be invested to pollution reduction, reduction of waste, increasing efficiency (of products, energy, production processes ...) and to looking for new sources of raw materials and energy.

Hot-dip galvanizing is an anti-corrosion system for steel that is fully recyclable and that reduces the need for using new materials. Its highly efficient and long-term protection from corrosion by means of galvanized zinc coatings makes the level of the physical useful life of steel products equal to their technical-economical service life and, at the same time, reduces the costs of routine maintenance. Emissions of pollutants from modern plants are at such a low level that the presence of this technology is not perceived by the population as negative or disturbing even in urban areas. Comparison of the energy demands of the production of non-ferrous metals clearly shows that the specific energy consumption for the production of zinc is the lowest of those commonly used non-ferrous metals. An important factor is also the fact that zinc is a fully recyclable metal. It can be regularly recycled without any

Hot-Dip Galvanizing of Steel Structures. DOI: http://dx.doi.org/10.1016/B978-0-08-100753-2.00014-8

change in its physical or chemical characteristics. At the same time, life cycle analyses indicate that the hot-dip galvanizing anti-corrosion system has a considerably lower environmental impact compared to the anti-corrosion protection system based on application of an organic coating material [55].

Hot-dip galvanizing could be considered as a process with so-called "soft sustainability." This means that the total economical value of the resources or products obtained will not be reduced in future. It admits that primary non-renewable resources can only be consumed if a respective equivalent is generated (i.e., the consumption is not loss-making). After the expiration of the useful life of a product obtained, the product must be completely recycled to avoid generation of loss.

Note

Strong sustainability is considered to be difficult to achieve both from the short-term and long-term point of view at present. It requires a non-decreasing volume of resources, which means that the strong sustainability principle only makes it possible to consume renewable resources as a source of energy or raw materials. It does not consider non-renewable resources at all.

Figure 14.1 Possibility of prefabrication and time saving on-site.

References

[1] MgA. Ester Havlová. Room in the landscape. Author's archive, 2012.

[2] Malouin PJ. <http://www.boulangerie.net/forums/files/3_Malouin_portrait_oval.jpg>, 2015.

[3] Volta A. <https://cs.wikipedia.org/wiki/Alessandro_Volta>, 2015.

[4] Galvani L. <http://commons.wikimedia.org/wiki/File:PSM_V41_D302_Luigi_Galvani.jpg>, 2015.

[5] Faraday M. <http://www.general-anaesthesia.com/images/michael-faraday.html>, 2015.

[6] Sorel S. <http://en.wikipedia.org/wiki/Stanislas_Sorel>, 2015.

[7] From the history of hot dip galvanizing. Tribotechnika 2013/1. <https://www.scribd.com/embeds/124908180/content?start_page=1&view_mode=book&access_key=key-1qntwavstuwtnlny98y7&show_recommendations=false>, 2015.

[8] Novák P. Types of corrosion. Corrosion and material protection, 2005/4.

[9] Bystrianský J. Corrosion in waters. Teaching text Faculty of Mechanical Engineering, Czech Technical University in Prague, 2009.

[10] Corrosion. <http://cs.wikipedia.org/wiki/Koroze>, 2015.

[11] Technical encyclopedia, SNTL Prague, 1982.

[12] Novák P, Kouřil M. <http://www.vscht.cz/met/stranky/vyuka/predmety/koroze_materialu_pro_restauratory/kadm/pdf/1_0.pdf> Electrochemical corrosion of metals. Institute of Metallic Materials and Corrosion Engineering, Teaching text University of Chemistry and Technology Prague, 2015.

[13] Pernice L. Zincatura a caldo. Roma: Associazione Italiana Zincatura, 2007.

[14] Metallic coatings deposited electrolytic and chemical. ČOS 999904, CZ Department of Defense, 2012.

[15] Rune T, Torgny W. Manual hot dip galvanizing. Czech revised edition CSGA, 2011.

[16] Zmrzlý M, Fiala J. Diffusion factors in system zinc-carbon steel. METAL, 2002.

[17] Ostrá V. Sherardizing and its possible applications. Faculty of Mechanical Engineering, Czech Technical University in Prague, 2008.

[18] Kuklík V. The roughness of the zinc coatings executed according to EN ISO 1461. KONSTRUKCE, 2011/6.

[19] Peissker P. Hot dipping of wire in molten zinc and zinc-aluminium-Alloys, 2002.

[20] Kania H. High-temperature galvanizing. Conference CSGA, 2008.

[21] Kudláček J, Kreibich V. The issue of cleaning surfaces. Faculty of Mechanical Engineering, Czech Technical University in Prague, 2008.

[22] Richtlinie VDI 2579 Emissionsminderung Feuerverzinkungsanlagen. Verein deutscher Ingenieure, 2008.

[23] Feldmann M, et al. Hot-dip-zinc-coating of prefabricated structural steel components. Luxembourg: Joint Research Centre of the European Commission, Publication Office of the European Union, 2010.

[24] Mec K. Regeneration of flux in the process of hot dip galvanizing. Conference CSGA, 2011.
[25] Kuklík V. Post on the issue of safety of steel structures of hot dip galvanized structural components, 40. Amsterdam: Procedia Engineering, 2012.
[26] Kuklík V. Selected examples rough structures of zinc coatings. Collective galvanizers, ČSPÚ, 2013.
[27] Kreibich V, Kudláček J. Negative effects caused by hydrogen at the process finishes. Progressive and innovative surface treatments. Center Surface Treatment, 2009.
[28] Bárta J. Hydrogen embrittlement of steels according to their strength characteristics. Corrosion and Material Protection, 2007.
[29] Pluhař J, Koritta, J. Engineering materials. SNTL/ALFA, 1977.
[30] Burton BP, Perrot P, Okamoto H. Phase diagrams of binary iron alloys. Materials Park, OH: ASM International, 1993.
[31] Su X, Tang N-Y, Toguri JM. Thermodynamic evaluation of the Fe-Zn systém. Amsterdam: Procedia Engineering, 2001.
[32] Zmrzlý M, Schneeweiss O, Houbaert Y, Fiala J. Phase transformations galvanized coatings observed in the Mössbauer spectra. METAL, 2005.
[33] Nakano J. Thermodynamics of the Zn-Fe-Al-X systém. McMaster University, 2005.
[34] Rajak P, Tewary U, Das S, Bhattacharya B, Chakraborti N. Phases in Zn-coated Fe analyzed through an evolutionary meta-model and multi-objective genetic algorithms. Amsterdam: Procedia Engineering, 2011.
[35] Liberski P, Podolski P, Kania H, Tatarek A. Theoretical and practical aspects of zinc coatings on ferroalloys created by hot dip galvanizing process. Konference AČSZ, 2008.
[36] Kuklík V. Measurement and evaluation of the thickness of the zinc coating using magnetic methods. Progressive and innovative surface treatments. Faculty of Mechanical Engineering, Czech Technical University in Prague, 2008.
[37] Sandelin RW. Galvanizing characteristics of different types of steel. American Hot Dip Galvanizers Association, 1940.
[38] Kuklík V. The appearance of hot dip galvanized coatings. Prague, Finishes. 2008/2.
[39] Maass P, Peissker P. Handbook of hot-dip galvanization. WILEY-VCH, 2008.
[40] Schulz WD, Thiele M. Feuerverzinken von Stückgut. Eugen G. Leuze Verlag, 2011.
[41] Arbeitsblätter Feuerverzinken. <http://www.feuerverzinken.com/wissen/arbeitshilfen/arbeitsblaetter/>, 2015.
[42] Katzung W, Schulz W-D. Beitrag zum Feuerverzinken von Stahlkonstruktionen-Ursachen und Lösungsvorschläge zum Problem der Rissbildung. Stahlbau, 2005/4.
[43] Horstmann D. Fehlererscheinungen beim Feuerverzinken. Verlag Stahleisen GmbH, 1983.
[44] Kuklík V. Coating on thermally cut surfaces. Konstrukce, 2011/6.
[45] Kuklík V. White rust on zinc coatings. Konstrukce, 2008/4.
[46] Kuklík V. Repairing defects zinc coating. Wiegel CZ, 2010.
[47] Kudláček J, Kreibich, V, Červený J. Limited resistance galvanized materials in hot tap water. Faculty of Mechanical Engineering, Czech Technical University in Prague, 2012.
[48] Kuklík V. Service life of zinc coatings. KONSTRUKCE, 2011/2.
[49] <https://www.blake-uk.com/content/12-galvanizing>, 2015.
[50] Kuklík V. Design and execution of components for hot-dip galvanizing. Tribotechnika, 2011.

[51] Kuklík V. Study of liquid metal assisted cracking at hot dip galvanizing. Faculty of Mechanical Engineering, Czech Technical University in Prague, 2013.
[52] Scanning electron microscope JEOL JSM-7600F. <http://www.chempoint.cz/videt znamena-vedet>.
[53] Línek J. Old peoples' home Hagibor. From author's archive. Prague, 2009.
[54] Woodley T. Galvanizing and sustainable construction. EGGA, 2008.
[55] Kudláček J, Kreibich V. Sustainability surface technology. Faculty of Mechanical Engineering, Czech Technical University in Prague, 2011.
[56] Pohribný J. Galvanizing plant. Author's archive. Prague, 2005.

Index

Note: Page numbers followed by "*f*" and "*t*" refer to figures and tables, respectively.

FSC
www.fsc.org
MIX
Papier | Fördert gute Waldnutzung
FSC® C083411

Kolibri 360 GmbH
Ferdinand-Jühlke-Straße 7
99095 Erfurt, Deutschland
produktsicherheit@kolibri360.de